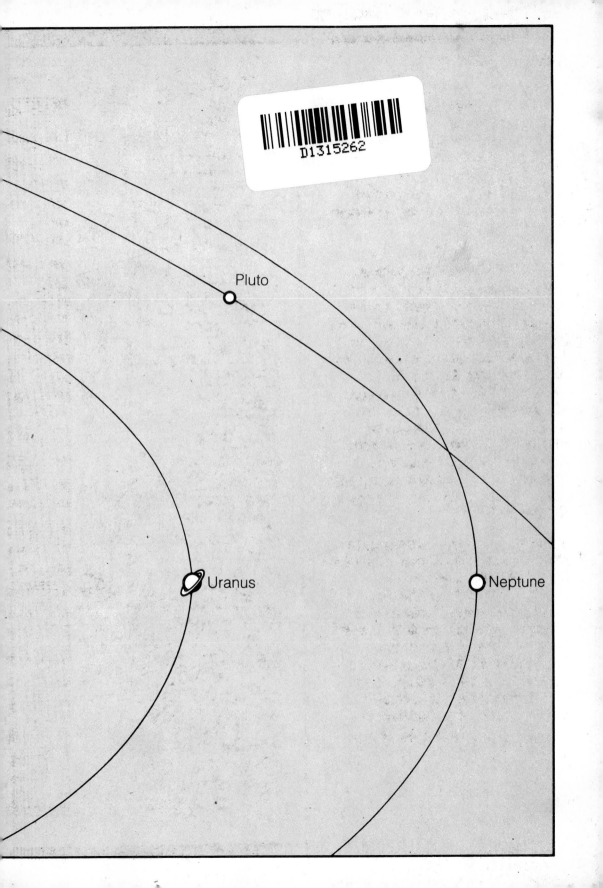

SPACE HISTORY

Also by Tony Osman

In Aid of Surgery
The Discovery of the Universe

SPACE HISTORY

Tony Osman

ST. MARTIN'S PRESS
NEW YORK

ISBN 0-312-74945-7

Library of Congress Catalog Card Number 83-50684

First published in 1983 in Great Britain by Michael Joseph.

First U.S. Edition
10 9 8 7 6 5 4 3 2 1

CONTENTS

LIST OF ILLUSTRATIONS

COLOUR ILLUSTRATIONS

1

DREAMS OF SPACE

Mankind made its first firm move towards putting people into space on 4 October 1957, when the Soviet Union launched Sputnik, the first ever artificial satellite, into orbit. Only three and a half years later – on 12 April 1961 – the technology had advanced so far that a man, Yuri Gagarin, could be put into orbit, sustained there, and then return to earth. And on 20 July 1969, two men, Neil Armstrong and Edwin Aldrin became the first people ever to set foot on any celestial body other than the earth. From Sputnik to the moon walk took less than twelve years.

This 'great leap for mankind', as Armstrong called it, was a triumph of an almost unimaginably complex technology. A powerful booster rocket had to be developed; the astronauts were required to master all the subtleties of space navigation; they needed the protection of an infallible life-support system, and a moonlander that could never be tested on earth, yet must never fail in space. But it would be misleading to talk as if the technologists behind the space effort simply saw a complex of problems and set about solving them: if that had been the case, the moon landing wouldn't have happened so quickly. For a couple of decades at least, they have, after all, been trying to devise a motor car that doesn't pollute the atmosphere, to control and harness nuclear fission, and to devise an economical supersonic aircraft.

Technological advances are not made simply because some momentum of history makes them inevitable, but because scientists and engineers want to make these particular developments and not others. The ambition and the knowledge are minimum requirements, and in addition, the scientists need resources, particularly money. The eight-year Apollo project to put a man on the moon is estimated to have cost the United States five thousand million dollars a year – and the researchers have to persuade those who hold the purse strings to loose them. Influencing committees, developing useful relationships with non-scientists, attracting the interest of the general public: these are all arduous tasks, and researchers who can maintain enough pressure to succeed are themselves inspired by more than just the interest of solving a scientific or technical problem.

However, the urge to put people onto the moon and after that establish settlements in space has a special appeal that converts scientists and technologists into fanatical enthusiasts who work on despite interruptions. For hundreds of years, man has been fascinated by the idea of travelling in space, and in particular to the moon; many space enthusiasts report how they were inspired initially by science fiction, particularly Jules Verne and H.G. Wells, both of whom wrote 'practical' science fiction.

The first known story of a trip to the moon was rather a long way from being 'practical' SF – not surprisingly because its author, Lucian of Samosata, lived and wrote in the second century AD. His book, *A True Story*, is literally a flight of fancy. Its hero sets sail from the Straits of Gibraltar, and intends to cross the Atlantic 'to see what is on the other side'. He is carried off by a storm that eventually deposits him on the moon, whose inhabitants are, at the time, fighting those of the sun. He joins with the moon people, who lose, so Lucian is carried off to the sun. He later returns to the moon as a result of a peace treaty. At the end of a fascinating and inventive story, he sails back to earth.

The first scientific inspiration for space stories came from the great scientific revolution of the sixteenth and seventeenth centuries, and from Galileo in particular, who looked at the moon and the planets through the telescope he had perfected for astronomy. Before Galileo, the Aristotelian view had prevailed: the planets and the stars were special objects, not subject to the laws of earthly physics, and the moon had an intermediate status. After Galileo, the heavenly bodies were perceived as being natural and normal, and appropriate places for a visit.

One of the earliest stories about moon travel was actually written by a great astronomer, Johannes Kepler: his *Dream about the Moon* of 1609 is accurate in its descriptions of the view of the earth from the moon, and on the strange pattern of day and night on the moon – fourteen 'days' of each.

The first story of moon travel which became a classic was *Man in the Moon* by Francis Godwin, who eventually became a bishop under Queen Elizabeth I. His hero, Domingo Gonsales, is towed to the moon by a team of 'gansas' – wild swans that he has tamed. He notices on the way that while the earth begins to look like another moon as he travels away from it, it is still distinguishable by the fact that its rotation is much quicker. The moon, as he approaches it, turns out to be mainly covered with water: only the patches that appear dark to us are land. On arrival he meets a group of moon people – they are much taller than humans and travel around by flapping feather fans. Godwin mentions that the moon has a weaker, smaller 'attractive power' than earth, so that people can leap and bound great distances. The political and moral systems described are Utopian.

In 1638 another Elizabethan bishop, John Wilkins, published a *Discovery of a New World* – the moon. He even suggested a range of possible ways of getting there, one of which involved harnessing birds and making a flying chariot. Considering that he was writing before Newton had evolved the idea of universal gravity, he had one striking insight: Wilkins reckoned that 'an Orb of Magnetic Vigor' encircled the earth, but that this got weaker as one travelled outwards. If a man could 'get 20,000 miles upwards, it were possible for him to reach the moon ultimately'.

Although these early romances of moon travel gradually came to reflect the growing knowledge of astronomy, they were vague about the question of transport to the moon – not surprisingly, as not even a steam engine had yet been invented. But rockets *had* – they had long been used in warfare. The first proposal for a rocket-powered trip to the moon was in Cyrano de Bergerac's *Voyage to the Moon* of 1657. He even proposed that the rockets be lit in bursts – an early form of multi-stage rocketry.

Other writers used their accounts of voyages to the moon as a basis for satire: the most distinguished of these was Daniel Defoe who, in 1705, published *The Consolidator*, or *Memoirs of Sundry Transactions from the World of the Moon*. Its aim was political and religious criticism – Defoe was a Nonconformist – but his account of the trip shows how effectively the heavens had been de-mystified in the century since Galileo. His spacecraft was a chariot balanced on two giant wings (with a total of 513 feathers, the number of Members of Parliament at the time). The wings were set flapping by a heat engine which was only rather vaguely described (the age of steam was just starting) and the craft *flew* towards the moon. Defoe realised that there was a point on the way where the attraction of the moon would overcome that of the earth, so that the wings would have to act as what are now called retro-rockets. He also believed that the moon was like the earth, with the same kind of inhabitants: or if he didn't believe it, it suited his purpose to say that he did. He showed some scientific humility by pointing out that the inhabitants of the moon would see earth as their moon, but his book, like others of the time, cannot be called scientific: they were all satires. They gradually gave way to the more realistic science fiction that inspired the scientists who eventually sent people to the moon.

The drive that led to manned space travel came from the nineteenth century and was a mixture of an urge to colonise and a pure love of science and its achievements. It was a time when laymen were beginning to be excited by new technological developments: it was a uniquely stimulating period in science because the scientific achievements were more accessible than they had been previously. Science became attractive and fashionable – for the first time London Society attended lectures at the Royal Institution. Alessandro Volta demonstrated the first electric battery, and Humphry Davy soon used the battery to discover new chemical elements. Michael Faraday, who succeeded Davy as Professor at the Royal Institution, invented the dynamo, the transformer and the electric motor. The electric telegraph came into being, and the first steamship crossed the Atlantic. Railway lines spread across continents, and adventurous people experimented with balloon travel. Science suddenly seemed omnipotent: very shortly nature would be fully understood. Whatever man wanted to do, he would achieve in the closely foreseeable future.

This was the spirit of the new science fiction: it was inspired by and closely based on current scientific developments. The early stories of Edgar Allen Poe fit this description exactly: his *The Unparalleled Adventure of One Hans Pfall*, was the first 'realistic' account of a moon trip. Poe himself was an enthusiastic amateur scientist, and his story is laced with scientific detail. He claimed that it was derived from Hans Pfall's own manuscript, dropped from a balloon on to a public square in Rotterdam. According to the message, the balloon had arrived from the moon seeking help for Hans Pfall, having transported him there in the first place. If one were prepared to accept this as possible – and there was then a widespread opinion that at least a tenuous atmosphere extended as far as the moon – then Hans Pfall's account was a simple extension of known scientific achievement.

His balloon was filled with a gas 'never yet generated by anyone but myself' with a density 'about 37·4 times less than that of hydrogen'. We now know that there is no such gas – hydrogen is the lightest possible element – but the idea was not purely fanciful in Poe's time. Hans Pfall carefully makes and varnishes his balloon and loads it with food and water, some scientific instruments including a compass, a telescope and a barometer, a couple of pigeons and a cat. He also takes an impossible, fictitious device, a machine that is described as being able to condense air. Poe used the word condense to mean concentrate – the device would take in the rarefied air that lay between earth and the moon, or so Poe thought, and put out normal, atmospheric air at atmospheric pressure. Hans Pfall had this to breathe.

Pfall knows the distance to the moon and that it varies as the moon travels around its elliptical orbit: it will take a minimum of 161 days to arrive there if he travels at the 60 mph that can be managed on earth. He appreciates that he has only to reach the null point – the Lagrange point as it has since become known – where the moon's gravity balances that of earth. His problem will then be to control his descent.

14

His ascent starts as planned. He uses a blast of gunpowder (which also eliminates his major creditors) as a booster and he ascends peacefully and predictably to a height of seventeen miles or so, noting as he goes the changing appearance of the heavens above and the earth below. By this time he is feeling some ill-effects from the lack of oxygen, and he decides to set up his condensing apparatus. There is a careful description of the way he fits a rubber bag around the basket of the balloon – the bag has three glass observation windows – and, once it has been fitted, Pfall pumps the air condenser and fills his 'spacecraft' with air dense enough to breathe. Such a condenser would, in fact, have exploded the rubber overcoat, but the story has the kind of scientific detail that makes the trip seem possible. It is a verisimilitude that inspires.

Hans Pfall's navigational method has the same plausibility. He intends to ascend until the moon's gravity pulls him in. To achieve this, he must stay beneath the moon: otherwise, he reasons, he will miss it and go on for ever. Fortunately, his balloon drifts until it is positioned approximately over the equator; he then ascends until the moon's gravity starts to take effect. Soon after this, the balloon inverts, and what has been an ascent from earth becomes a descent on to the moon. As Pfall descends he notices flares from erupting volcanoes, but he soon becomes more preoccupied with survival. He has to throw his ballast and even his scientific instruments overboard so that he can land at a safe speed.

When he does land, he is quickly surrounded by moon creatures: dwarfs without ears, because speech is, as Poe knows, impossible in the rarefied atmosphere of the moon: we are promised further details in a later letter. Pfall's first report is said to be delivered to earth by one of these dwarfs. It is only at the end of his account that Poe suggests that the voyage was imaginary and mentions that a dwarf from a travelling circus has been missing for several days.

Although this was the story of a trip that never occurred, there was, Poe makes clear, no reason why it should not occur in the future. In his notes at the end of the book, he says that his story of a moon voyage is different from any that had been written before: it is 'an attempt at verisimilitude, in the application of scientific principles ... to the actual passage between the earth and the moon'.

Perhaps because Hans Pfall's story was admittedly a hoax, or because technology was not yet looking so far afield, Poe's book did not greatly influence the development of ideas about space travel. One book that did was Jules Verne's *De la Terre à la Lune* (*From the Earth to the Moon*), first published in 1866. It describes what was intended originally as an unmanned moonshot with a spacecraft that was, rather late in the planning, converted to take three passengers.

Jules Verne's method is more practicable than Poe's. The whole

15

adventure is originated by a group of American cannon-makers left with time and production capacity on their hands by the ending of the American Civil War, and the spacecraft is sent off by a giant cannon. The book has the characteristic verve of the best nineteenth-century writers and also embodies their philosophy – an unbounded enthusiasm for science and an uncritical colonialism. It is interesting to note at this point that colonialism has been a consistent motif of space exploration until very recently. The first men to land on the moon planted an American flag there, and the books and interviews of the time spoke of moving to other planets when our own was overcrowded or overexploited. It is only recently that space researchers and propagandists, taking account of modern political thinking, have started referring to 'space islands' rather than 'colonies'.

Verne's prospective space traveller, Michael Ardan, has the task of explaining the project to the group of cannon-makers. He has an inspiring view of the essential simplicity of the project. As he says, the distance from the earth to the moon is a little less than nine circumferences of the earth, and 'there's no sailor or experienced traveller who hasn't covered a greater distance than that in his life'. He points out that it would take an express train only 300 days to reach the moon. Then follows one of the prophecies that at the time so excited the imaginations of scientists: 'I don't think I am going too far in saying that, in the near future, there will be trains of projectiles in which people will be able to travel comfortably from the earth to the moon.' The journey would, in fact, be more comfortable than rail travel at the time. Then Michael Ardan is asked whether there is life on the moon? Considering the enormous adaptability of life forms, he cannot see why there shouldn't be.

It is easy to understand why scientists, and the general public, found Verne so stimulating. The essential practicality of his vision reflected the nineteenth-century optimism that inspired, say, Brunel to build steamships of undreamt-of size. The spirit of the time decreed that if a task was possible – if it did not actually contradict the laws of nature – then it would be achieved. Verne is full of details that certainly seemed at the time to be realistic. His fundamental idea of launching the spacecraft from a cannon has a weakness that he could barely have suspected (the astronauts would be killed by the enormous acceleration), but in his writing there were all sorts of details that testified to his scientific knowledge. The distance of the moon from the earth and the speed of the missile are used to calculate the correct time of 'blast-off' which will ensure that the missile intercepts the moon. The cannon is to be pointed directly upwards, so it must be placed under the moon's path – the chosen site is Florida, which was also the launching site for the successful Apollo flight to the moon. And, because Verne knew that the moon returns to any given location in the sky only once in rather more than eighteen years, his characters recognise what modern scientists call a 'window' – a short, unique period for a successful launch. The landing on the moon is to be cushioned by firing rockets: the reaction would slow the spacecraft. This idea anticipates the retro-rockets that are actually used by modern spacecraft. Verne was aware of the need for oxygen on the journey and proposed to make it by heating potassium chlorate; he also suggested that the carbon dioxide in the exhaled air should be removed by passing the air over potash – still one of the standard methods used. He couldn't, however, predict the food that twentieth-century astronauts would consume: instead, his crew stocked up with tinned food and compressed vegetables in the manner of long-distance mariners.

Jules Verne wrote two 'moon books'. At the end of the first, the spacecraft is still travelling: to the observers at a super-telescope on earth, it looks as if it will miss the moon. In the second, *Round the Moon*, published four years later in 1870, we find out that the spacecraft did just that, but subsequently returned safely to earth. *Around the Moon* (*Autour de la Lune*) is written from the point of view of the astronauts and takes up the story much earlier than the end of the previous book, at launch-time, the moment of blast-off. If anything, it has an even more practical tone than its predecessor. The astronauts, for example, lie down during blast-off to minimise the effect of what are known as g-forces, the downwards forces felt by the astronauts as a result of the vigorous upwards acceleration of their spacecraft. A couple of dogs that the travellers had, in typically Victorian fashion, taken with them were exempt from these precautions and one of them dies of g-force injuries. He is eventually 'buried in space' as sailors are sometimes buried at sea. The buried dog then accompanies the spacecraft throughout the rest of the trip until the retro-rockets are fired. (This sequence parallels the real-life experiences of the astronauts in Apollo 13. That craft was damaged by an explosion on its way to the moon and some of the debris from the explosion accompanied the ailing vehicle throughout its flight. At one point there was so much wreckage that the astronauts had great difficulty getting navigational 'sights' with a sextant.) Verne's passengers were so inspired by their discovery – that any object placed outside the craft would automatically accompany it – that they realised the possibility of spacewalks. The provision of an air supply was the only remaining problem.

Jules Verne's flight nearly ends in disaster when the astronauts find themselves on what seems to be a collision course with a meteor. They miss it, but pass so close that their path is slightly deflected by its gravitational effect: eventually they realise that this deviation explains why they missed the moon. For Verne it was just as well that they did

17

miss, as he hadn't devised a way of taking off again. His travellers loop around the moon and, as a result of the combined gravitational effects of the moon and the earth, return to a successful splashdown in the Pacific Ocean. This idea, that a spacecraft could return automatically from a trip around the moon, without any need for an extra impulse en route, was essential if spacecraft were to be hurled upwards by cannon. It is typical of Verne's assurance when handling detail. He didn't foresee every eventuality, nor solve every problem, but the way he described his machines and their use made space travel seem eminently attainable.

The other major science-fiction writer was of course H. G. Wells. He describes a moonflight in *The First Men on the Moon*, published in 1901. Wells was not only enthusiastic about the popularisation of science; like Verne, he also had political interests. His ideas eventually crystallised into a belief that the world would be better run by a technocracy.

This is not the spirit of *The First Men on the Moon*, which is ostensibly written by an unsuccessful entrepreneur named Bedford who had, somewhat improbably, turned to playwriting as a way of making money. While at Lympne, working on his plays, he meets a Mr Cavor who has invented a substance – Cavorite – that nullifies the force of gravity. Inspired by the commercial possibilities of the material, they go into partnership: they decide to cover a spherical spacecraft with Cavorite arranged on roller blinds. By manipulating these, they can make the spacecraft accessible to the gravitational force of only chosen celestial bodies. If they open the blind that faces the moon, they will travel towards it. They equip the spacecraft with food, drink and the cylinders of air they will need for the trip, then they set off.

This basic idea is, even now, scientifically plausible. We can fairly easily make shields that cut off magnetic or electrical forces; it is not certain that the gravitational force is so different that it cannot be counteracted. At the beginning of the twentieth century the idea was even more acceptable, and Wells, like Verne, studded his story with scientific details that made it extremely persuasive. He described the way that the travellers would take 'compressed foods and concentrated essences' for nourishment, and cylinders of compressed oxygen for respiration, together with 'arrangement for removing carbonic acid and waste from the air and restoring oxygen by means of sodium peroxide'.

The voyage was undertaken partly in the spirit of scientific enquiry, partly as a colonising expedition, and partly as sheer adventure. This last aspect of the story, which still appeals to readers, develops on the moon. The explorers land on a dark area and notice at dawn that the advancing sunshine is melting what turn out to be 'icebergs of air'. This melting permitted the travellers to abandon their spacecraft with its life-support system, because the moon, it turned out, develops a respirable atmosphere, although the increasing heat of the sunshine eventually makes it uninhabitably hot. It turns out, though, that it is unbearably hot on the surface only, and the astronauts find trapdoors that take them down to the subterranean world of the Selenites.

There are various races of Selenite, and some of them are highly intelligent. They look rather like large – five-foot high – insects and they live underground, some of them emerging at night to herd the giant mooncalves that form the basis of one part of their economy. Their only structural material is gold, which makes the moon as attractive to the astronauts as America was to the Spaniards.

The moon in Wells's book is a world turned inside out, with a network of chambers and passages – the homes and cities of the Selenites – and the sea at the centre. The underground passages are lit by streams of a phosphorescent liquid that flow towards the sea. While the two astronauts are exploring the passages, they eat an intoxicating toadstool, are captured by Selenites while under its influence and imprisoned underground.

The adventure quickens. They escape, fight their way to the surface, and are then faced by the problem of finding their spacecraft before the sunlight destroys them. They are considerably hampered by plants which unfurl and grow to enormous sizes on what had seemed at first to be a barren plain. Eventually the entrepreneur astronaut finds the craft, but when he sets off to collect Cavor, he discovers that his scientist colleague has been re-captured by the Selenites. Bedford takes off on his own, and after some deft manipulation of gravitational forces he makes a safe splashdown near Folkestone. He takes some sample gold bars ashore and would have made a return journey to the moon had not his craft been sent into space by a small boy who tampered with the Cavorite blinds.

Bedford settles down to capitalise on his voyage by writing an account of it, and is stunned to hear that a Mr Weedigee is receiving electromagnetic messages 'similar to those used by Signor Marconi for his wireless telegraphy' from Cavor, now living with the Selenites on the moon. These messages describe the various races of Selenite and the structure and organisation of the Selenite society, and they bring the book to a fascinating close.

The theories of Poe, Verne and Wells were ingenious and exciting, but, as it happened, none of these writers predicted the method of propulsion that was eventually to be used to take man into space. Yet they inspired the people who, in the early years of the twentieth century, did achieve this feat. If one person is to be named as the father of space travel, it must be the Russian Konstantin Tsiolkowski, born near Moscow in 1857. Tsiolkowski was a brilliant theoretical engineer, but he

below, *Model based on
Tsiolkowski's design for a
rocket ship*

„Невозможное сегодня станет воз-
можным завтра." *Циолковский*

РАКЕТА К.Э. Циолковского

would not have turned to the problems of space travel if he had not been inspired by Jules Verne. In his writings he readily acknowledges this spiritual debt, but his understanding of rocketry and space travel owes nothing to any predecessor.

Tsiolkowski was the first to realise that the propulsion for space travel must come from rockets. For fully a thousand years, rockets had been used as playthings and weapons and had even been mentioned in the odd pre-scientific fable about moon travel, but their potential had not previously been appreciated. Tsiolkowski's rockets were practical devices or, to be more precise, they were designs for practical devices, for he never actually built one. But he understood, and noted in his diary as early as 1883, that a gas escaping into space would drive its containing vessel away from it. This is the principle of reaction flight. Tsiolkowski had discovered that the thrust of a rocket resulted from the ejection of gases: a rocket would, therefore, work in a vacuum. The thrust did not depend on having an atmosphere for the escaping gases to thrust against, and the propulsion would continue to be effective after the rocket had left the atmosphere that surrounds the earth.

Tsiolkowski's first publication (1903) was an amazingly detailed article called 'Exploration of Space with Reactive Devices'. Although he was a theoretician, not a rocket-builder, the discussions were strictly concerned with actual masses, forces and chemicals. It was in this article that he proposed liquid fuel for rockets. The alternatives are solid fuels, as used in toy fireworks, some military rockets and now the space shuttle, but, as Tsiolkowski realised, it is more difficult to control the burning of solid fuels. The development of manned space travel depended on liquid-fuelled rockets: the military could use solid-fuelled rockets. This difference sometimes caused problems for its space enthusiasts.

Tsiolkowski proposed using multi-stage rockets to solve the basic problem of space travel: the departure from earth. Any space rocket would accelerate itself as well as the spacecraft and, even though the mass of fuel would diminish during the flight, the mass of the rocket and the machinery it contained would remain constant. By Tsiolkowski's calculations, the mass of fuel would be four times that of the rocket and payload. There were enormous difficulties in carrying so much fuel in a single rocket, simply because the fuel had to be contained: the more fuel there was, the greater would be the mass of the container. The solution was to use a number of rockets, burning one after the other, each dropping away once it had burned out – a multi-stage rocket. This would make it relatively easy to overcome the force of the earth's gravity, and rockets, Tsiolkowski foresaw, would 'navigate interplanetary space, interstellar space . . . visit planets . . . or any other celestial body and return to earth'. He wrote a book about lunar exploration, *On the Moon*, which was published in 1935, the year of his death. The cover shows a couple of intrepid lunar explorers gazing at a lunar cliff, with the earth in the background. They are wearing the traditional uniform of explorers including a peaked cap, kneeboots and a knapsack.

Practical experiments with rockets during Tsiolkowski's lifetime were, in Russia, limited to firing rocket motors that were attached to sleds. Tsiolkowski was not involved in any of the developments in the liquid-fuelled rockets that were essential to space travel, but during this period the American scientist, Robert H. Goddard, was developing ideas very similar to those of Tsiolkowski and managing to put them into practice. Goddard was born in Massachusetts in 1882 and, while he is on record as being stirred by the science fiction stories of H. G. Wells, his devotion to space travel arose from a near-mystical experience. He tells the story in his autobiography which was first published modestly, long after it was written, in the journal *Astronautic* in 1959. At the age of seventeen he climbed a cherry tree to trim branches and happened to look at the countryside below him. He was amazed by the beauty of the view, and realised how wonderful it would be to make some device in which he could ascend towards the planets, watching the earth recede as he went. He relates that he was a 'different boy when he descended the ladder. Life had now a purpose for me'. This vision sustained him through both technical difficulties and public ridicule. The experience was so important to him that, throughout his life, he observed the anniversary of the day it occurred.

Goddard's first idea was to project his spacecraft by what he called a 'centrifugal drive' – whirling the craft around a horizontal shaft and eventually hurling it into space. He manufactured a range of models to test the idea, but eventually came to the conclusion that Newton's laws of motion held: a slingshot launch would not work for a spacecraft. He realised that rockets could be used, but he wanted to avoid them because of their expense, 'unless there is certainly no other way'. In about 1909 he set about designing a space rocket. He certainly didn't know about Tsiolkowski's work but he reached the same conclusions: he recognised that the ideal combustion mixture was of oxygen and hydrogen and that a multi-stage rocket was required to overcome the earth's gravity.

Goddard needed money for his research and, shortly before the start of the First World War, he produced a report, 'A Method of Reaching Extreme Altitudes', that persuaded the Smithsonian Institution in the USA that he was worth backing. During the war he diverted his rocket research into producing weapons and, for these relatively crude applications, a solid-fuelled rocket was adequate. He returned to liquid-fuelled rockets after the war.

His report was published as a book in 1919, and it contained notes on the amount of fuel required to raise 'one pound to an infinite altitude'. This was, of course, a calculation of the fuel needed to take a projectile into space and on to the moon. At this time Goddard was thinking only of unmanned exploration, but he did suggest that his projectile should contain flash powder arranged in such a way that it would detonate when it struck the moon. This proposal, the only scientific way of knowing that his rocket had landed on the moon, was so dramatic that it put Goddard in the public eye overnight – a 'moon man' – and, as the public was not really prepared for such an advanced idea, he was ridiculed. He had to allow the excitement to die down before he could return to his research. There were very few others working on rocket travel at the time, and Goddard had virtually no contact with any of them. After 1923, when he severed his connections with military work, he returned to his university in Massachusetts and the solo work that was so congenial to his temperament.

Goddard's researches covered every aspect of rocketry. One important technological milestone was his launching on 16 March 1926 of the world's first liquid-fuelled rocket. As far as his own programme was concerned, another milestone was his meeting with Colonel Charles Lindbergh, the Atlantic flier, during November 1929. Lindbergh was a

Dr Robert H. Goddard beside his first successful liquid-propelled rocket, 16 March 1926

man of influence, and he persuaded both the Guggenheim Fund and the Carnegie Institute to put money into Goddard's research.

Goddard needed somewhere more open than Massachusetts for the rocket development that now became possible, and he established himself at the Mescalere Ranch, near Roswell, New Mexico. He stayed there until 1941, quietly improving his equipment. He produced liquid-fuelled rockets that reached speeds of 700 mph and heights of 8000 feet; he studied the use of a gyroscope to steady his rockets, and ways of assembling rockets in clusters to increase the total thrust; and he developed the idea of sending payloads of instruments into space for research. He returned to government work, again on military rockets, in 1941 and died on 10 August 1945. Although he had been spurred in his research by the idea of space travel, and had for a while been famous for advocating it, by the time of his death he had disappeared from the public eye. The immense originality and inventiveness of his work has been truly appreciated only since the dawn of the space age.

*Dr Robert H. Goddard
and colleagues at
Roswell, New Mexico.
Goddard is holding
cap and pilot
parachute of his
improved rocket,
19 May 1937*

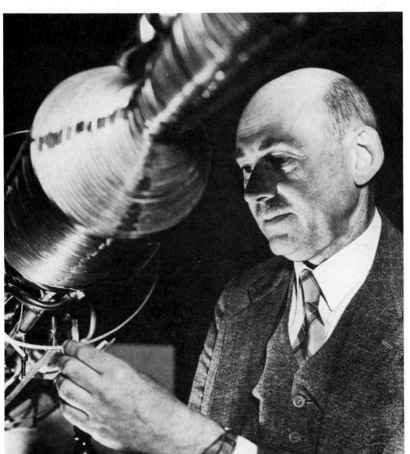

*Goddard making
adjustments to the
upper end of the rocket
combustion chamber at
Roswell, New Mexico,
1940*

Hermann Oberth, the third scientific forerunner of the space age, did achieve fame and recognition in his lifetime, partly because he sought it – he was a tireless propagandist for space research and space travel – and partly because he lived long enough to see men walk on the moon and unmanned cameras take pictures of Mars.

Born in 1894 in Hermannstadt, which is now part of Hungary, Oberth was yet another pioneer whose interest in space travel came from the science fiction writers, particularly Jules Verne. At the age of eleven he spotted that anyone fired from Verne's cannon would be killed by the acceleration. His own proposals were, at first, at least as eccentric. He suggested using a magnetic cannon, for example, and, like Goddard, a gigantic wheel that would spin the travellers into space. That he eventually turned to rockets was again due to Verne, who had proposed them as retro-rockets for landing.

Oberth's research was deflected into military rocketry during the First World War, but his interest in space travel was rekindled when he read a newspaper report of Goddard's 1919 proposal. Four years later he published his own first report: *Die Rakete zu den Planeten Räumen* (*The Rocket into Planetary Space*). This was only a tiny booklet, but it considered every problem of space travel – even its effect on the human body.

Now the stage was set. The nineteenth-century enthusiasm for space exploration had grown and it had inspired both theorists and daring engineers. The technology of space exploration was understood, but the achievement depended on the development of rockets of unprecedented power – rockets that could be assembled into multi-stage systems to take first instruments, then animals, and finally man into space.

What space exploration now waited on was an enthusiast capable of raising the vast funds needed for further research, by using arguments that were at best ingenious, at worst devious. But enthusiasm, even backed with money, would not be enough. The enthusiast also had to be a superb engineer, capable of designing and building multi-stage rockets in all their complexity.

The combination of all these talents would be rare. Yet two people with all these qualities appeared in the 1930s – Wernher von Braun in Germany, and Sergei Korolyev in Russia. They would be rivals until Korolyev's death.

2

THE ENTHUSIASTS

Money for space research has always either been public money, or military money on occasions when the space enthusiasts have persuaded the military that building rockets and going into space might be useful in war. When the money is public money, as it largely has been in the United States, the enthusiasts must communicate to the public the excitement and thrill of space travel.

Even though Great Britain has never managed to produce a rocket for manned space flight, members of the British Interplanetary Society are passionate advocates of space travel. The BIS was founded in 1933 by an amateur enthusiast, Philip Cleator, with a total membership of six people. The tally rose only slowly, though the society did manage to publish an occasional journal. Though no doubt at least some of the members had read Jules Verne and H. G. Wells, the BIS was not initially inspired by British science fiction of which there was very little at the time.

Before the war the BIS was always a small body, meeting in cafés and pubs, producing its journal on a shoestring. But it had influence. In 1935, Cleator published his book, *Rockets through Space* – an optimistic, inspiring book, written in a popular style and explaining in a rather general but essentially practical way how the problems of space travel could be overcome. It was successful: it sold well and it attracted new members to the BIS. One of these was Arthur C. Clarke, the most influential British science-fiction writer of his period, and one whose books are always based on sound, realistic physics.

But though the society's publications were serious, and its small membership excited by developments, Britain as a nation was not moved towards space research. Arthur C. Clarke calculated that only a few hundred pounds were spent on rockets and space before the war. This was partly because of the size of the country, the lack of open spaces near the big cities, and the rigid laws preventing people from testing rockets, but also partly because British genius has never produced a rocket engineer.

The American equivalent of the BIS, the American Interplanetary Society, had different origins and a different destination. It was founded in 1930 by David Lasser, editor of *Wonder Stories*, a science-fiction magazine, and the original members were largely science-fiction writers – indeed, many of them were contributors to Lasser's magazine. Lasser, like Cleator, wrote a popular book on space travel – *The Conquest of Space* – which he published in 1931.

Shortly afterwards, the society organised what turned out to be a gigantic meeting, addressed by the French rocket designer, Robert Esnault-Pelterie, and including a showing of a 'practical' German space film, *Frau im Mond* (*Woman on the Moon*). This meeting helped raise considerable public enthusiasm. Then Edward Pendray, historian and founder member of the American Interplanetary Society, went to Germany, met members of the essentially practical *Verein für Raumschiffs-fahrt* (Society for Space Travel), and agreed that the two societies should co-operate. This decision shaped the future of the AIS. It abandoned its space-travel romanticism and concentrated on hardware. It turned to rocket-building and testing, and changed its name to the American Rocket Society (ARS).

The beginnings of the ARS's rocketry were simple but ingenious. Pendray, his wife and one other member built a seven-foot, liquid-fuelled test rocket for a total of just over $30.00. Though the fuel was modern enough – petrol and liquid oxygen – the rocket itself was adapted from domestic hardware such as saucepans and cocktail shakers. It was fired, fairly successfully, in November 1932. A second version, fired the following year, went 250 feet up into the air, then it exploded. The society went on to build other rockets, many of them using a modern method of cooling in which the fuels cooled the combustion chamber on their way to combustion. This wasn't actually a new idea, but the previous inventors had kept the work secret, so that ARS's member James Wyld had to develop it for himself. His discovery made him and the society famous, and four members of its experimental committee set themselves up as a manufacturing company, Reaction Motors. This eventually became part of Thiokol Chemical Company, which built the liquid-fuelled rocket engines for the Bell X-1 – the first aircraft to travel faster than sound.

However, the AIS/ARS was no more progressive than its British counterpart when it came to getting men into space. This was left to the enthusiasts from Russia and Germany. The Russian effort was based on Tsiolkowski's investigations: he worked out the details of the designs that were needed and the thrust that would have to be developed but achieved little in practical terms. Very soon, however, the military programmes of the Soviet Union were demanding real, working rockets.

28

Military rockets have a long history. There is some doubt about exactly when they were first used because historians refer, somewhat ambiguously, to weapons such as 'fire arrows', and these may or may not have been self-propelled. However, the records from twelfth-century China confirm that the Chinese used rockets in war. In 1258 the Mongols used rockets to capture Baghdad, and as a result of this the Arabs learned how to use rockets.

The knowledge of gunpowder and of gunpowder-powered rockets was probably brought to Europe by the Arabs or the Mongols, and the weapons are clearly described by Marchus Graecus during the thirteenth century. In the fourteenth century an Italian, Muratori, actually coined the word *rocchetta*.

Although the Italians certainly used rockets in war in the Middle Ages, and the French used them in the defence of Orléans in 1429, and soon after that at Bordeaux and Gand, they were only a minor part of the armaments used. Their serious military use in Europe started at the end of the eighteenth century and resulted from two savage attacks on the British by rocket-firing troops under Tippoo Sultaun of Mysore during the two battles of Seringapatam in 1792 and 1799.

This led to an astonishing period of British military rocketry, developed by a British engineer of genius, William Congreve, who was eventually knighted in recognition of his achievements. He designed large (32-lb) rockets with a range of some 3000 yards to be used in the Napoleonic Wars. In 1806 a barrage of 2000 of these rockets was discharged from small boats in an attack on Boulogne, and in 1807 25,000 of them were fired against Copenhagen.

Congreve-type rockets were used throughout the nineteenth century – in the war against America of 1812, and in the American Civil War; and they were even employed in the First World War by the French against German Zeppelins and balloons – military rockets worked well as incendiary devices. Otherwise, their main use by this time was for signalling or for illuminating battlefields.

All these rockets were like, or developed from, the domestic firework: they used solid explosives to propel them. Once this solid propellant was lit, in no matter how sophisticated a form, the rocketeer had to retire immediately. The combustion of a solid-fuel rocket cannot be controlled. This is a perfectly suitable system for a military rocket, where the energy is needed simply to hurl the projectile forward. And, because a solid-fuel rocket can be built and left standing, ready to fire, for years, solid fuels are ideal for defence, provided enough energy is available.

The chief disadvantage of a solid-fuel rocket is that the amount of thrust produced by a given weight of fuel is relatively low. A rocket using liquid fuels, such as petrol and liquid oxygen, kerosene and liquid oxygen, or liquid hydrogen and liquid oxygen, can produce much more power from fuel of the same weight. In addition, a liquid-fuelled rocket is controllable. With an admittedly complex arrangement of containers and valves, the thrust can be increased, decreased, or changed in direction. A liquid-fuelled rocket can be started and stopped more or less at will. This kind of motor is ideal for space exploration, but it has one great disadvantage for military purposes: these rockets have to be stored empty and fuelled when the need arises. This was a serious problem for the space enthusiasts of Russia and Germany, who needed to persuade their governments of the benefits of financing the development of liquid-fuelled rockets.

The development of Russian rockets is shrouded in a mixture of secrecy and fiction, but it is clear that the government did back the development of rockets from around 1930, and that there were two principal research groups involved. The Leningrad Gas Dynamics Laboratory (GDI) was formed in 1928 to develop solid-fuelled rockets for military use. By 1930 it had designed an anti-aircraft rocket and was on the way to an anti-tank rocket. Around this time it started to look at liquid-fuelled rockets, and it industriously produced and tested dozens of small, liquid-fuelled rockets, continuing its work, to no great avail, until the end of the Second World War.

The society that played the greatest part in developing Russia's spacecraft - in giving that nation a stream of 'space firsts' - was the Group for the Study of Jet Propulsion, known in Russia as GIRD. A Moscow branch of GIRD was founded in 1931 and, although it was an official body principally concerned with civil defence, it attracted space-travel enthusiasts. When the Soviet government amalgamated the Leningrad and Moscow GIRDs to form the State Reaction Scientific Research Institute in 1932, they appointed as leader the brilliant engineer Sergei Korolyev, the designer whose energy and vigour was behind all the early Soviet space successes. Rocket propulsion and space travel were lively topics in the Soviet Union at this time. The branches of GIRD had translated and published the work of various foreign enthusiasts and scientists, but in fact Russia's own output was on an infinitely vaster scale. The years from 1928 to 1932 saw the publication of the nine volumes of an enormous encyclopedia on space travel. *Interplanetary Communications*. Though this had, not surprisingly, only a small circulation, a single-volume book, *Interplanetary Travels*, by Jakov I. Perelman, one of the founders of the Leningrad branch of GIRD, sold 150,000 copies.

The enormous public interest provided the drive and the researchers; the government's desire for weapons guaranteed funding; and the genius of Korolyev in particular, and of those who worked with him, all com-

bined to produce the rockets and spacecraft with which Russia shot into the lead in the Space Race.

The most vigorous pre-war rocket research, and the widest spread, came from Germany. It was engineers from this country who eventually put men on the moon, but because Germany had lost a war, it was Americans who were the first there. However, the rocket that took them to the moon was a linear descendant of pre-war German rockets, and the progress of the US space programme was accelerated by a German engineer, Wernher von Braun.

Germany's space society was the famous *Verein für Raumschiffsfahrt –* (*VfR*), founded in 1927. Max Valier, one of its members, persuaded Fritz von Opel to try and build a rocket-propelled racing car. Wernher von Braun, as a young man, imitated these experiments with a model racing car driven by a firework.

Von Braun's mother was an amateur astronomer who introduced him to the futuristic books of H. G. Wells and Jules Verne and gave him an astronomical telescope. He was a rather slack, undisciplined schoolboy until he came across a copy of Oberth's great book, *By Rocket to Outer Space*. Then he discovered that he needed a lot of mathematics to follow this exhilarating subject and became a keen student.

Sergei Korolyev (second from right) with other GIRD colleagues in the summer of 1934

31

Curiously, it was Germany's position after defeat in the First World War that gave von Braun the backing he needed to achieve his dreams. Even so, he needed some agility in argument to keep the funds flowing. The Treaty of Versailles prohibited Germany from having a military establishment: she was not permitted an air force, was not allowed to build tanks, and could not develop heavy artillery. Any nation would have tried to find a way round these restrictions, but as Germany was a military nation with a powerful and important military caste, she found the prohibition particularly restrictive. In order to circumvent the demands of the treaty, the Germans set up a gas warfare training-school in Russia, another one for the use of tanks, and began to develop a rocket force as an alternative to heavy artillery.

Professor Karl Becker, who was head of the ballistics and armaments branch of the German Army, asked an engineer, Walter Dornberger, to develop, secretly of course, a solid-fuelled rocket system for short-range attacks, and one using liquid-fuelled rockets to carry bigger than ever loads of explosives further than the range of any known gun. The Germans were not going merely to do as well without howitzers as they might have done with them: they intended to do better. Karl Becker must have been under the influence of space enthusiasts even to have considered developing liquid-fuelled rockets, but the recommendation was crucial. As a result of it, the German Army produced the V-2 military rockets, whose descendants were eventually to take men to the moon.

It was Rudolf Nebel, the head of *VfR*, who had asked Wernher von Braun to convince the army that it made good military sense to encourage the development of rockets. Von Braun went directly to Karl Becker and spoke both as a member of the German ruling classes and as a serious engineer. He didn't mention space exploration, and came across to Becker and Dornberger as the only member of the *VfR* not 'filled with baseless enthusiasm'. At first sight the *VfR* had seemed to Dornberger to be a society of romantics.

In 1932 Dornberger built a test-stand for a liquid-fuelled rocket, and von Braun built a small rocket to be tested. The success of the test firing enlisted Dornberger into the rocket movement, and he quickly became a space enthusiast. It was an encouraging start. Von Braun enlisted the best of his colleagues from the *VfR* to help in his rocket research: the association had run out of funds during the savage German depression and closed down. But its aims lived on in the German Army, and the engineers found the opportunity to work on their space rockets while they were being paid to develop rockets for military use. The Dornberger/von Braun group certainly was not concentrating solely on military rockets, but equally certainly they did develop the V-2s, which were

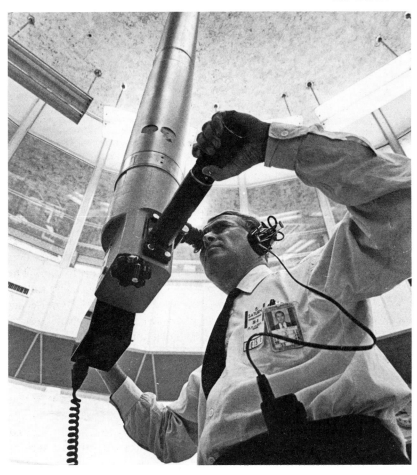

used in the Second World War and provided the basis for today's intercontinental missiles.

The Dornberger group started their research at Kummersdorf, a suburb of Berlin, but that was clearly not a good place for building rockets and it was a thoroughly unsuitable one for testing them. In 1935 the group decided to move. Von Braun remembered a holiday resort that he had enjoyed, an isolated peninsula near the Baltic Sea called Peenemunde, secluded enough for secrecy and remote enough for safe rocket testing. The rocket team started to build a new centre there, but the move took years: it wasn't complete until August 1939. Needing enormous amounts of money, yet working in a Germany that was deep in an economic depression, Dornberger, von Braun and their eventually eighty-strong team needed considerable resourcefulness.

33

In 1934, two liquid-fuelled rockets (code-named A-2) flew success-fully. Von Braun numbered his rockets consecutively – A-1, A-2 and so on – but when the rockets were used as weapons they were renumbered to give them a dramatic title. The A-4 rocket became the V-2 missile (the V stood for the German word for vengeance). Now the team had results to prove that they could build long-distance rockets, and they started to seek funds and help from various military sources. At times, at least, they seem to have realised that the air force and the army, for example, were rivals: they could play one off against the other to their advantage. (Though technically still banned, the German air force was openly growing – by 1935 it equalled the British.) Von Braun convinced the man in charge of aircraft production, von Richtofen (cousin of the famous Red Baron), that a rocket-propelled fighter plane would be immensely valuable to the air force, and von Richtofen offered the money to develop one. But von Braun was already in the pay of the army and his ultimate chief, Karl Becker, offered more money for research, partly to decrease the significance of bomber planes. Eventually, the army contributed 300 million Reichsmarks which was to be used to build and equip Peenemunde.

Von Braun had to use his persuasive powers again at the end of September 1939. Germany had invaded Poland, and Britain and France had declared war on her. The nation needed to use everyone to their best advantage, and there was doubt that the work at Peenemunde was sufficiently valuable – Hitler himself didn't think much of the project at that time. So von Braun held an Open Day, invited dozens of academics – professors of ballistics and engineering; experts on telemetry and gyro-control – and told them what his needs were, making it clear that their departments at the universities would receive research money from the government, and that they themselves would be more useful where they were than in active military service. The scientists and engineers offered their support.

Eventually, the team had to convince Hitler that the rockets could be valuable weapons, an exercise which it is fascinating to consider in retrospect because von Braun and Dornberger have both since said that the rockets weren't developed as weapons, and that they engineered them as such only because there was no other way of raising the necessary funds – certainly they were expensive and inefficient weapons. But by 1943 the Germans had failed in their attempt to beat Britain by using long-range bombers, and now Britain, helped by its invention of radar, was gradually winning control of German airspace. So von Braun and Dornberger showed Hitler a film demonstrating A-4 rocket flights – the first had been in October 1942 – and pointed out that these rockets would travel above the atmosphere at enormous speeds, invulnerable to any defence. Hitler was persuaded: the team was told to develop the A-4 rockets as the V-2 weapon, and a factory was set up in tunnels under the Harz mountains at Mittelwerk. In the end, a rather modest 6200 V-2 missiles were built. Some 3000 of these fell on targets, others failed, and around 2000 were still in storage at the end of the war.

By the beginning of 1945 von Braun, along with many of his country-men, had recognised that Germany would lose the war, and that the victors would be Russia, America, France and Britain. After discussion, they decided that it would be best to be captured by the Americans: they were anxious to continue with their research on space travel, so needed a large, rich nation. Russia, though large and rich, was an ominous police state.

Peenemunde, with its proximity to the Russian forces, was awkwardly placed for their purposes, so in February, the senior technicians stated that they needed to be near Mittelwerk to continue the fight. Five thousand people, innumerable wagonloads of equipment and tons of research documents were all moved, with only a few problems, through a country which was in the last stages of losing a war. Near Mittelwerk they buried tons of plans and other documents, then settled down to wait for the Americans: the Harz mountains were directly in the path of the US advance. When the Germans finally surrendered, von Braun set out what amounted to a prospectus, telling his captors how useful his team could be to them. The Americans were only partly convinced, but when they realised that the other Allies were raiding Germany for technicians, patents, inventions and equipment they mounted 'Operation Paperclip' and took an assortment of technical experts to America, including the cream of Peenemunde, just over a hundred members of staff. Then followed the documents, the plans and around a hundred completed rockets. This was the start of the Space Race, though no one knew it at the time.

The Russians also recruited people from Peenemunde: although the best experts had gone to the USA, a number of useful technicians and scientists still remained, including Helmut Grottrup, one of von Braun's assistants. There were enough people to re-start the production of V-2s, which were obviously attractive weapons to Russia, a nation surrounded by potential enemies and American bases. Stalin encouraged the develop-ment of both long-range space bombers and intercontinental missiles, and Sergei Korolyev was put in charge of research. Russia set out to build a range of gigantic rockets, and it was one of these that took Sputnik into orbit on 4 October 1957.

3

SPACE AGE LIFT-OFF

The launch of Sputnik, the Russian word for traveller, took the United States completely by surprise: the immediate reaction was one of shock and astonishment. Russian technology had not hitherto attracted much attention. In the USA motor cars were common and sophisticated; television was widespread; agriculture mechanised and effective. Americans had invented the atom bomb and devised fast aircraft. Surely they could claim world leadership in technology: yet in 1957 they weren't even certain exactly how to put up an artificial satellite. Some American and British scientists went on record as saying that the Soviets couldn't have done it; that they were somehow fooling the world. But most people believed that it had happened, and found that their confidence in American pioneering spirit and technological expertise were at best open to question.

The second reaction was fear. In the midst of the Cold War Russia had alarmed the West with the threat to Berlin and the takeover of Hungary, and was then a lively threat to other Western nations. It had developed an atom bomb – some said that it had stolen Western secrets to do so. And where traditional warfare was concerned, Russia had an immense superiority in numbers over the West. Until 1957 people had felt that this superiority was counterpoised by the American superiority in scientific know-how, but the orbit of Sputnik made that argument doubtful. Though the scientists were sceptical, American popular imagination held that bombs could be dropped from orbiting spacecraft; the country was alarmed and wanted action.

What the Soviets had, and the USA wouldn't have for a very long while, was a gigantic, powerful rocket. The Russians had made plans for this soon after the end of the Second World War when they began to see the United States as their chief enemy – a country thousands of miles away from the industrialised west of the Soviet Union, out of the reach of Soviet aircraft. In 1947 Stalin proposed an intercontinental missile, based on the German V-2 rocket. 'Such a rocket could change the fate

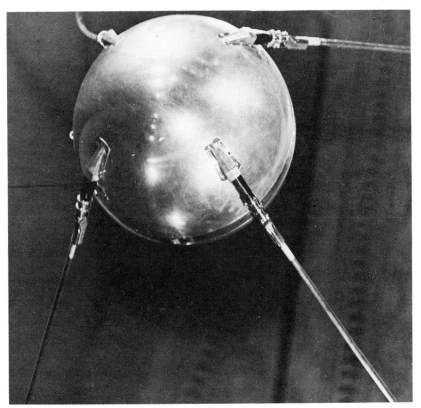

Sputnik 1, the first Soviet artificial earth satellite launched 4 October 1957

of the war,' he said. 'It would be an effective straitjacket for that noisy shopkeeper, Harry Truman.' The Russians started producing V-2s – more briskly than Germany had ever done. Quite quickly their V-2s were also more powerful than the German models, although their range was still limited to 500 miles or so.

The awesome intercontinental missiles, designed by Korolyev, came into being under Khrushchev. In 1956 Korolyev developed a giant missile, the R-7; he wanted to use it to launch a satellite, but was told that it was not a valuable project. He therefore continued work on the R-7 as a military project and, some time in the summer of 1957, he achieved a successful firing – over 4000 miles – of a rocket that could carry a two-ton bomb.

There was a snag. Russia could now use the new missile in international bargaining, but while it was one thing to mention the achievement, it was another to be believed. Khrushchev suddenly realised that an orbiting satellite would be a very good demonstration of the potential of the rocket, and Korolyev's romantic space dreams at last had an attractive practical application. He was told to put up a satellite, and quickly. It took only six weeks to assemble the multi-stage rocket and put aboard a stunningly simple satellite – not much more than a battery-powered radio transmitter. It was successfully launched in the evening of 4 October 1957.

The West had previously been so convinced of Russia's technical backwardness that it ignored announcements that could have prepared it for Sputnik. In October 1956 Moscow news had discussed a possible satellite. On 10 June 1957 the Soviet Union had told the committee of the International Geophysical Year that its satellite was ready. In September it had announced the frequency Sputnik would use for broadcasting. And in August the Soviet Union reported its successful firing of an intercontinental ballistic missile (ICBM). America didn't believe any of this, simply because it didn't want to believe.

The Americans themselves had turned away from developing ICBMs. They had successfully produced long-range bombers to deliver atom bombs, and the bombs were much too big to be fitted on to any rocket they envisaged. Why not stick to developing bombers, which they understood? Scientific experts agreed. Dr Vannevar Bush, one of the inventors of the computer, and one of those who had urged Roosevelt to make an atom bomb, told a Congressional committee in 1946 that 'There has been a great deal said about a 3000-mile high-angle rocket. In my view, such a thing is impossible today and will be for many years ...' Charles Wilson, Defense Secretary to President Eisenhower, was, like Eisenhower, opposed to research into powerful rockets. To get backing, such research would have to show an 'immediate military need'. President Eisenhower misjudged both the significance of Sputnik and the mood of the nation: he referred to Sputnik as, 'One small ball in the air, something that does not raise my apprehensions, not one iota'. For support he called on White House aides – Sherman Adams, who stated that the USA was not trying to win 'an outer-space basketball game', and Clarence Randall, who called Sputnik 'a silly bauble ... a bubble in the sky'. Randall was reassured by Queen Elizabeth and Prince Philip, who visited Eisenhower soon after Sputnik's flight and said, he reported, that 'People in London gave it one day of excitement, and then went about their business.'

The American public was less sanguine. It wanted to know how the US had got so far behind in the technology race, and it wanted to see an American satellite in orbit. The problem was not that the USA did not have a programme to design rocket missiles: on the contrary, it had rather too many and it even had a candidate for a space rocket. The invention of the H-bomb – far smaller and lighter than the atom bomb – in 1953 had meant that long-distance rockets could carry a significant warhead. As a result, five different rockets were under development and any one of them could have been used as part of a satellite-launching system. But the nation's efforts in this multiplicity of projects were dissipated by internal squabbling and by the fact that, at that stage, only one of the projects had been accepted by the government as a space rocket, as opposed to a missile launcher. This was the Viking, originally designed as a sounding rocket – one to study the atmosphere – and very difficult to develop.

One severe problem was rivalry between the three American services: it was clear that whichever of them developed a space rocket, once there was an H-bomb to carry, would become the dominant service. The army proposed Project Orbiter, which would use the Redstone rocket it had under development and would be organised by Wernher von Braun. The air force suggested a programme based on the Atlas, an ICBM started in 1953. The navy put forward what was essentially a civilian programme, Project Vanguard, based on its civilian research rocket, the Viking. This had been chosen by Eisenhower, on 29 July 1955, as the means of launching unmanned satellites for the International Geophysical Year. Predictably there was a certain amount of wheeling and dealing among the contenders, and the air force was eventually persuaded to back the navy's project. Donald Quarles, then Assistant Secretary for Research and Development for the Department of Defense, supported this alignment, which suited Eisenhower's very odd idea of keeping separate the development of military and research rockets. Project Vanguard was chosen, and then – perhaps because it was 'merely' a civilian project – starved of money. And as American rocketry's greatest asset, the Wernher von Braun team, stayed with the army, Vanguard lacked the talent it needed.

Even so, by late 1957 Vanguard did manage to produce a three-stage rocket with a tiny, $3\frac{1}{2}$-lb satellite that the government unwisely billed as the nation's reply to Sputnik – which weighed 184 lb. The publicity this claim attracted was so enormous that a launch had to be made, so on 6 December, at 11.44 a.m., after a series of delays the rocket was fired. It rose a few feet, crashed and exploded. The rumour of American inadequacy seemed to be confirmed when the Russians said how sorry they were about the Americans' difficulties. The Russians were speaking from a position of great strength: a month earlier, on 3 November 1957, they had launched Sputnik 2, a half-ton satellite that carried a living creature, the dog Laika, with its life-support system.

Von Braun's determination to put satellites into space had been unaffected by the fact that the US government had asked someone else to

Left, *Laika*, and right, *Sputnik 2, the Soviet satellite which took Laika into space, 3 November 1957.*

do so. He had established a base at Huntsville, Alabama, after the end of the war, and with the help of army chiefs, who had resented the choice of Vanguard, he had continued to test the Redstone rocket, allegedly as part of the research on the intermediate-range Jupiter. He needed to design a nose cone that would survive re-entry from space so he modified a Redstone with extra stages, calling the product a Jupiter-C. In September 1956, a year before Sputnik, he successfully fired the first Jupiter-C 600 miles into space and 3000 miles across the earth. This was so obviously a potential satellite-launcher that the Pentagon instructed von Braun and the army that they must not on any account launch a satellite. One possible reason is that Eisenhower didn't want to interfere with Project Vanguard.

Following the launch of Sputnik 1, the US government saw things in a somewhat different light. The army and von Braun were told that Project Explorer, using the Jupiter-C, would be very acceptable indeed. On 31 January 1958 the Jupiter-C, with its Explorer satellite aboard, took off and duly went into orbit. The USA had started badly, and late, but now, at last, it was a runner in the space race.

Immediately they could claim a first: Explorer 1 made a startling space

40

discovery. One of its scientific instruments was designed to measure and count electrically charged atoms in space. The scientist running the experiment, James van Allen, noted that the counter seemed to block, then he realised that it had been overloaded. It had run into a completely unexpected belt in space, thick with electrically charged atoms and smaller particles. He is now immortalised in the name given to such clumps: the van Allen belts.

There was no doubt in the mind of anyone concerned that there *was* a space race and that the runners were up and going. The American public was seeking a spectacular success to quell their fears of the technological prowess of the Soviet Union, which was thought to reflect their military prowess.

Vanguard eventually succeeded in putting its satellite into orbit on 17 March 1958, but that hardly constituted a stride forwards or even a hopeful indication for the future. A couple of months later, on 15 May, the Russians launched Sputnik 3, weighing a ton and a half, and of this nearly a ton represented scientific instruments. The USSR was clearly finalising its preparation to put a man into space.

Inspired by the romance and challenge, to say nothing of the prestige represented by a man in orbit, US public opinion demanded a manned moon rocket. The demand was strong enough to be heeded, but the way ahead wasn't clear. Eisenhower, by now old and ill, was still not totally convinced of the threat of Sputnik and the developing Russian space

Lift-off of Explorer 1

41

programme, although he eventually conceded to public opinion, perhaps illogically, by agreeing to mount an American programme for putting a man into space. But he was slow moving – too slow for the space enthusiasts and too slow for the military departments who said, and may well have believed, that the defence of the USA depended on being able to put a man into space, and even on being able to establish military bases on the moon – an idea that is still bizarre but that certainly wasn't recognised as such at the time. The military had as their ally the Democratic senator for Texas, a long-time space enthusiast named Lyndon Baines Johnson. In the wake of Sputnik 2 he had already organised an enquiry into the problems and future of the US space effort. Eisenhower, on the other hand, wanted to restrict the influence of the 'military-industrial complex', an alliance composed of the military who wanted to spend the nation's money on weapons and the big industrial companies who were delighted to take the money. He wanted this group kept out of space research; although it was the spokesman for the military who were generating the pressure to put a man into space and making it clear that it was a defence project, worthy of lavish spending.

The navy's Vanguard programme had no momentum. Vanguard 2 had been launched with only a tiny satellite, and out of a total of eleven attempts to launch Vanguards, there were just three successes. The project could not be developed to take a man into space, and the navy had no other. The army and the air force both had missile programmes and programmes for putting a man into space. The air force referred to their programme as 'quick and dirty' and called it 'Man in Space Soonest' which had the worrying acronym of MISS. This was to be a rushed attempt to put a man on the moon, but had little scientific validity. It was to develop from the flights of the 'X' series of supersonic planes which were going faster and faster and higher and higher, and would eventually go into space, though not into orbit.

The air force also had a second programme called Dyna Soar which was organised by Walter Dornberger. This involved a spacecraft roughly resembling an aircraft which would be taken by rocket or jet power into orbit and would then return to land like a plane. This principle was later developed as the Space Shuttle, but the Dyno Soar programme itself was cancelled in December 1963, at a time when there was a strong move to limit the number of approaches to space exploration. In its early form the project would have put a man in space, but there was no way of developing it to take men to the moon, which was, by then, the aim of the USA.

The army also had its man-in-space programme, and, in addition, Wernher von Braun had already started designing a very powerful rocket, the Saturn, for military purposes. Thus both services showed great ingenuity in demonstrating why putting people into space was a military project and as such required financial backing from the state.

Eisenhower and his scientific adviser, James Killian, the president of the Massachusetts Institute of Technology, were impressed by the arguments of scientists and by the strength of public opinion in favour of a space effort, but they were anxious to keep the effort in civilian hands and well away from those of the military, who tended to be spendthrift and had no genuine scientific purpose in view. They needed to find, or create, a civilian body to direct the space effort.

It turned out that there was already one in existence, the National Advisory Council of Aeronautics, which had been set up in 1915 and had a good record in aeronautical research. On 2 April 1958 Eisenhower and Killian retitled this modest organisation the National Aeronautics and Space Agency (later the National Aeronautics and Space Administration) and charged it with peaceful and scientific space exploration. The acronym NASA was to pass into history.

The Act establishing NASA was signed in July 1958 and work was underway a couple of months later: NASA was entrusted with putting a man into space and with the moon-survey projects that both the army and air force had been working on. As a result of the organisation and rationalisation of the last few years the air force had been responsible for the development of military boosters: now it was decided that NASA should be responsible for civil ones. It was given control of the Jet Propulsion Laboratory in Pasadena, but it lacked an engineer. Eventually Wernher von Braun and his team were recruited and it is a fair guess that von Braun, who had always been a space-travel enthusiast, helped to make that decision, even while persuading the military to back his research.

By the time that von Braun joined NASA at the end of 1959 work had started on Project Mercury, which was to put the first American in space. This involved developing the Mercury spacecraft to withstand the shocks of take-off and re-entry, with a single orbit in between. Seven astronauts were assigned to the project, and six of them eventually went into space. The seventh was found to have a heart problem and was taken off the list of spacemen, though he continued as co-ordinator. All these early astronauts eventually became famous and rich.

Project Mercury was to include space flights by animals – mice first and then monkeys – and eventually a flight by a robot. Following that there was to be a sub-orbital flight by a man, and finally a real manned space flight and an orbit.

While the Mercury series was under way, NASA was developing its moon-landing programme which it code-named Apollo. Congress wanted high priority given to putting a man on the moon (and getting

43

him back) in the 1960s, and provided enthusiastic support for the elaborate and expensive series of flights that was proposed. President Eisenhower, however, was less encouraging, having been led to believe by his scientific adviser that the main attractions of putting people on to the moon were 'emotional compulsions and national aspirations'. He cut NASA's 1962 budget.

In January 1961 John Kennedy was sworn in as President of the United States. He was extremely interested in emotions and national aspirations and had already attacked Eisenhower for allowing 'a missile gap' to develop between the USA and the Soviet Union. He, and indeed the nation, by then saw Eisenhower as a deadening force: Kennedy stood for 'the new frontier', for revivification. He would clearly be prepared to back a manned space programme, but only if the prospects looked good: he didn't intend to back a failure. To begin with, they looked terrible. In its first three years NASA had tried to put twenty-eight unmanned satellites into orbit and only eight had succeeded. The first Mercury launch failed after fifty-eight seconds; the second rose only four feet from the pad. Later flights were more successful, but progress was slow.

Meanwhile, in 1958 and 1959 Russia had launched Lunas 1, 2 and 3. The last of these, launched on 7 October 1959, had travelled round

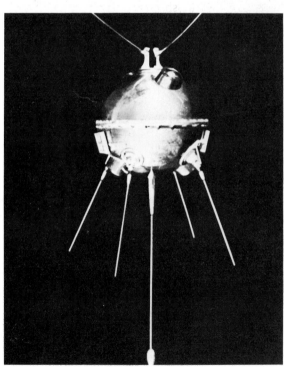

Luna 1 automatic probe, the world's first artificial solar satellite, launched on 2 January 1959

Luna 3, the Soviet interplanetary automatic probe which was the first to fly around the moon and to photograph its far side

behind the moon and taken photographs. On 20 August 1960 the run of Soviet firsts continued with the successful recovery of live dogs from orbit – Sputnik 5. This was the first time that a living creature had returned from space, and the achievement suggested that it would not be long before there was a Soviet man in space.

Then, at last, the Americans started to notch up successes, if not firsts: on 31 January 1961 the Mercury project successfully lobbed Ham, a chimpanzee, into space. This was not intended to be an orbital flight, but it was a launch into and a recovery from space. Ham had worked his way through all sorts of tasks while aloft, and it would be possible to put a man into a similar sub-orbital trip. Alan Shepard was chosen as the first man in space, and the launch was fixed for the middle of March 1961.

Wernher von Braun, about to achieve a lifetime's ambition, behaved with due scientific caution. Ham's flight had not been perfect – he had travelled rather further than expected. Von Braun decided that the

45

*Ham, the chimpanzee
who was successfully
launched into space
aboard Mercury on 31
January 1961*

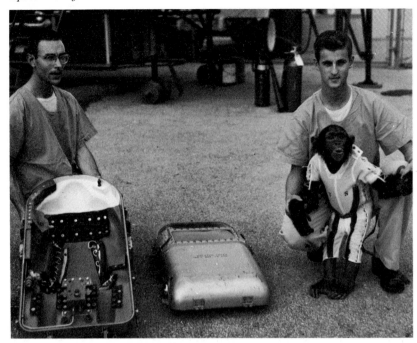

booster, the well-tried Redstone – a development of his V-2 – needed further testing, and he rescheduled the launch for early May. Alan Shepard duly went into space, reached a speed of more than 5000 miles an hour, and landed about 300 miles away from his launch site of Cape Canaveral in Florida. But von Braun's caution had been calamitous: the Americans had been beaten again.

In the meantime there had been another Russian launch, this time with a man aboard. On 12 April Yuri Gagarin had become the first human being to travel in space: he had completed an orbit and been recovered from a spacecraft that had come down by parachute in a target area on the Soviet landmass. Like the first satellite, the first man in space was a Russian.

It would be easy, though not completely accurate, to say that the Americans invaded the moon because they had failed in a military invasion. In 1961 President Kennedy was persuaded by his Central Intelligence Agency that the Cubans were ready to rebel against their President, Fidel Castro, and that a landing would be welcomed and successful. None of these statements was true and the Americans were soundly defeated in the Bay of Pigs in April 1961. This was just three months after he was sworn in as President. Kennedy needed a popular crusade that would restore America's belief in itself, in its image in the eyes of the world and, far from incidentally, in his own ability.

Manned space exploration seemed a promising area, but winning was important. The most obvious, striking and romantic challenge was to send men to the moon. NASA, from its inception, had had moon-trip enthusiasts and had made exploratory studies; it was sure that the USA could win. The Russians had achieved a lot of firsts, all using the same big rocket – a converted ICBM – as a booster, but going to the moon was quite a different project. Transporting a life-support system and an astronaut crew that far, landing them and getting them back would demand a very much more powerful rocket – about ten times as powerful. Wernher von Braun was already at work on such a rocket – the Saturn – and while the Russians might also be developing a more powerful booster, there was no certainty of this. The Americans were confident that if they put enough money into Saturn, they could outdistance the Russians. The organisation of technology was exactly what they were good at.

The project attracted Kennedy: a moon-landing would be a splendid realisation of the phrase which had characterised his election campaign, 'the new frontier'. And NASA's advice seemed more reliable than that of the CIA. In his address to Congress on 25 May 1961, Kennedy made his historic declaration: 'I believe that this Nation should commit itself to achieving the goal, before the decade is out, of landing a man on the moon and returning him safely to earth.' He backed his statement with a large increase in NASA's funds, and the project was named Apollo.

They seemed at the time to be brave words. Even though the USA hadn't yet put a man into orbit, and its unmanned space research had started badly, NASA had committed the nation to the most daring enterprise imaginable. However, it made technological sense. One might imagine that if men can be put on the moon in a decade, a similar development in the field of medicine should produce a cure for cancer or for heart disease. But a trip to the moon was not dependent on new scientific discoveries: in 1961, and indeed long before it, we knew that men walking on the moon would need space suits, a special spacecraft for landing, and highly advanced communications and navigational systems. The spacecraft would have to be very safe, with back-up equipment wherever there was a risk of breakdown. It would need life-support systems – atmosphere and temperature control to provide protection from the extremely high temperature that would be produced when the spacecraft, returning at a very high speed indeed, struck the earth's atmosphere. And to launch all this into space it would need a very powerful booster – a rocket more powerful than any built before. Yet nothing here was completely novel. It was all a matter of development.

NASA was already well funded, and the romance and challenge of space travel was attracting some of the best technologists. Their programme involved making longer space trips, testing more developed spacecraft until the moon landing became an obvious, safe next stage. Each new development was to be tested on the ground, then in space, and finally with a crew.

The next stage was a repeat of Alan Shepard's trip. On 21 July 1961 Virgil Grissom made a sub-orbital flight and landed, like Shepard, at sea. (The Mercury capsule was too small, and its booster too weak, to carry the machinery that would have been required for it to come down, as the Russians did, on land.) The plan was that the capsule would drop into the Atlantic, where it would float until rescue helicopters and boats arrived. Then the astronaut would operate a switch that blew off the bolts securing the hatch. He would climb out and be carried to the safety of an aircraft carrier.

What actually happened is that the bolts blew and the hatch opened long before the rescue parties were ready. Grissom just managed to scramble out before the tiny spacecraft sank. As a result, NASA decided that hatches should in future be more difficult to open. Ironically, and tragically, this decision helped to make one of the Apollo spacecraft a death-trap in which Grissom and two others were to die on 27 January 1967.

Grissom's first flight was a success and the Americans could consider going further, but the Russians were making progress too. On 6 August 1961 Herman Titov took off to spend a day in space. He was very sick as he became weightless but he ate meals in space and even slept there. It was clear that man could live in space. Titov's trip marked yet another step forward for Soviet space research, another political advantage at the time when the Berlin Wall was being built. The Russians were in need of good propaganda.

It was not until the following year that the Americans achieved even one orbit, but on 20 February 1962 John Glenn satisfactorily completed three. An Atlas D Booster took his craft, Friendship 7, into orbit between 100 and 163 miles up. Glenn reported the splendid sights of space and successfully counteracted the effect of a malfunctioning control jet.

Then disaster looked imminent. In order to recover the spacecraft, it had to be slowed by firing a retro-rocket which would allow it to drop back out of orbit and hit the atmosphere. It was fitted with a heatshield, that would melt away and thus keep the spacecraft cool.

Before the retro-rocket was fired, NASA officials on the ground discovered that the heatshield had been dislodged. The risks were appalling. If the heatshield was knocked aside as the spacecraft hit the atmosphere, John Glenn would be incinerated. Equally, he couldn't simply

left, *Launch of Friendship 7, the spacecraft in which John Glenn successfully completed three orbits in space, 20 January 1962*

below, *John Glenn lifts himself into Friendship 7 during a pre-launch checkout*

stay in space. NASA worked out an emergency approach very quickly: Friendship 7 was in its third orbit and was due to re-enter at the end of it. Normally the retro-rocket, once fired, would be discarded, but as it was right in the centre of the part of Friendship 7 that would strike the atmosphere, Glenn was told to retain it and use it as a heatshield. He did this and re-entered perfectly. Once he had landed, it turned out that the heatshield was in any case perfect and that it was the instruments monitoring its performance which had failed.

Though modest – only three orbits – and though troubled by what was in fact an instrument failure, Glenn's trip into space was a success; it was taken as a good augury for the moon race. Glenn himself became a national hero.

The Americans had always been open about their achievements: both their successes and their failures were reported on radio and television. When NASA worried about Glenn, the public knew about it. The Russians, on the other hand, were secretive. They announced their space flights at best while they were occurring, but often only after they had ended, and they didn't record any failures. Not until much later did they release any pictures of their rockets or spacecraft, and they even invented a launch site, called Baikonur cosmodrome, which was actually a couple of hundred miles away from the real town called Baikonur.

The result of this paranoiac secrecy was that rumours were circulated, suggesting that the Russians had had disasters and concealed them, or even that they were so keen to win the race to the moon that they were sacrificing astronauts recklessly. None of this was actually true – manned orbital flights would easily have been detected from earth by their signals – but it did mean the Russian announcements were often greeted sceptically. It was also pointless for them to lie about Baikonur: there are few things a spy satellite can spot more easily than a space-launch site.

There followed a period of gentle consolidation and extension. On 11 August 1962 the Russian Andrian Micolaev went into space for four days, setting a new record for the longest flight. Pavel Popovich went up a day later into much the same orbit. This was an achievement in itself – putting up two spacecraft on successive days was something the USA couldn't yet attempt. And the obvious question was asked: why were there two spacecraft? In particular, did they intend to link? It is now certain that they didn't. Both were rather unsophisticated craft, and weren't able to manoeuvre in the delicate way that would enable a link-up. In due course, the two craft landed almost simultaneously.

Space watchers have since appreciated the different objectives that were involved in this particular Soviet exercise: Korolyev, as a cautious enthusiast, wanted to establish a firm series of toeholds in space. Gradually the trips were to become longer – a day would be followed by four days, and that by a week. Khrushchev, however, wanted headline-making sensations, so they compromised by agreeing that Korolyev would be backed for his four-day flight if he was prepared to put another spaceman up at the same time. Next, when he wanted backing for a week-long flight, he would have also to put a woman into orbit.

Though there was no reason against putting a Soviet woman into space, Korolyev and his group hadn't been considering it. There weren't any women astronauts in training; there weren't even any women jet aircraft pilots. But there were women doctors and engineers in the space programme, and these seemed obvious candidates. Khrushchev though, wanted an 'ordinary Soviet woman', and those in charge of the space programme had to find him one. After Gagarin's first flight there had been a flood of letters from people who wanted to become astronauts: now, in late 1961, the letters from women, which had been stacked away without any sorting, were produced and examined. They needed to find a woman who met Khrushchev's political criteria, and who could be trained as an astronaut in about eighteen months.

In the end, four were chosen. We know the name of only the one, Valentina Tereshkova, who was eventually selected to go into space. She was, at that time, twenty-four years old, a textile worker, a free-fall parachutist, and an active member of the Young Communist League – clearly a suitable candidate. On 16 June 1963 she went up for a planned 48-hour orbital trip that in fact lasted just under three days.

Tereshkova's flight created a sensation. Today women are readily accepted as equals in most spheres, but at that time it was a dramatic development. It seemed to demonstrate to the world the great advantages of the Soviet society, where everyone was equal and opportunities were open to all. Women in the United States began to wonder, out loud, why the US space programme involved no women astronauts. Tereshkova found that she had a new career: she became an ambassador representing to the women of the world how much better communist society was for them. She even married an astronaut, Nikolaev, who had been one of her instructors. Their wedding was a grand affair, and Khrushchev himself gave the bride away. It is difficult to confirm the reports that Tereshkova and Nikolaev have not lived together since Khrushchev fell from power.

While Tereshkova's flight was not part of the developing Soviet space programme, Bykovsky's was. He had taken off a couple of days before and stayed up for five days, thus setting a new duration record. Progress was slow but steady.

Meanwhile the Americans were continuing with the Mercury series, their one-man space flights. On 2 October 1962, Walt Schirra made a modest but apparently very successful nine-hour flight, although the problems he experienced on landing indicated another potential snag of space travel. When he stood up his face became pale, his legs purplish and swollen, and he felt a bit giddy. Normally the veins and arteries of the legs, and the heart itself, work to stop the blood from pooling in the lower part of the body under the effect of gravity. When there is no gravity, this system atrophies and, in Schirra's case, the results of nine hours in space proved very uncomfortable. He recovered, however, by the next day.

Gordon Cooper's trip, which started on 15 May 1963, was the last of the Mercury series. It was still of a modest duration – just over 34 hours – but it was used to demonstrate the value of a manned spacecraft. He spun and rolled the craft, he took an enormous number of photographs, and he made quite extraordinarily detailed observations of earth. From more than 100 miles up he claimed to see rivers, roads and even smoke coming from chimneys. Later astronauts were able to confirm the authenticity of this dramatic assertion.

These Mercury flights were by now near-routine, and the crucial public interest was waning. NASA needed a new programme. The one-man spacecraft were of a simple design: they couldn't be manoeuvred in space, and they weren't really suitable for a long stay. A trip to the moon and back would take just over a week and required a spacecraft capable of delicate manoeuvring, of making link-ups between, for example, the main spacecraft and the space-dinghy that would be landed on the moon. It would also have to be possible for the astronauts to make trips outside the spacecraft – on landing on the moon, of course, and perhaps on the way. So the Gemini series of two-man spacecraft was planned.

The Soviet one-man Vostok series finished at much the same time as the Mercury series – the end of 1963 – and Korolyev had already laid plans for his next spacecraft, the Soyuz, which would be a good deal more advanced than the Gemini. It would not be ready, however, until 1965 – and this didn't suit Khrushchev at all. Gemini was certainly very much more advanced than the Vostok craft, and it would take pairs of people into space: Korolyev had better put up a three-man craft. This might have seemed an absurd proposition but, if Korolyev wanted the funds for his more serious development projects, he had to find a solution for this problem.

On 12 October 1964, while the Gemini craft were still being developed, the world was stunned by the news that a new kind of spacecraft, the Voskhod 1, had gone into orbit, carrying a crew of three. Only one of these, Vladimir Komarov, was an astronaut: the other two were a doctor and an engineer. They went up in 'shirt-sleeves' – space suits weren't even carried. The Americans were chastened, but simply went on, logically and systematically developing the Gemini craft. The world accepted that the Russians were ahead in the space race. *Pravda* ran a headline: 'Sorry Apollo'. The fact that the spacecraft came back to earth after only one day raised a few questions but no doubts.

The propaganda victory masked the disturbing truth behind the daring venture: the Voskhod wasn't a new spacecraft at all, it was a crudely adapted Vostok. Korolyev had simply taken out the reserve parachute and the ejection seat. If the booster had failed – gone off prematurely, or misfired and gone off at an angle – the three crewmen would have been doomed. As they couldn't all sit on the one ejection seat, there was a certain grim logic about removing even that. The three men were crammed into the craft like three people in the front seats of a sports car – the third more or less on the shoulders of the other two. Once they were weightless it was relatively comfortable, but naturally there was no room for space suits, only for a day's food and oxygen, with reserves enough for a second emergency day. As the crew would have to land in the spacecraft – Gagarin and his successors had parachuted clear for the final stage of the descent – a small rocket was fitted underneath, designed to cushion the end of the descent. The Russian booster rocket was powerful enough to launch this technological monstrosity, and the world admired the apparent triumph.

Soviet engineers accepted the applause, no doubt a bit wryly, but the triumph did nothing for Khrushchev. The announcement that he had been replaced by Brezhnev followed closely behind the announcement that Voskhod 1 had landed safely. However, his spirit lingered long enough to inspire another 'Voskhod' flight and another Soviet first, a space walk.

Voskhod 2, modified for safety, could take only two people because it was necessary to reserve room for space suits and the equipment the astronauts needed, and because it was fitted with an airlock through which an astronaut could go into space without allowing the air in the spacecraft to leak away. And, a mere hour into orbit, Alexei Leonov did just that. He put on a back-pack with an oxygen supply, crept down the tunnel of the airlock and pushed himself out into space, secured only by a lifeline. It was a moment of exhilaration and triumph. The US Gemini launch was due in five days' time, but the Americans would have to be content to achieve only the third multi-person space launch – with only two people – and the second space walk.

Yet, while they talked of triumphs, the Russian reports concealed problems. The space walk came near to calamity. The pressure inside the spacecraft was normal atmospheric pressure – 14 lb per square inch. In space, a suit at that pressure would be pumped up so hard that it would be rigid, but the pressure inside it couldn't be too low, because the astronaut would then risk the 'bends' – the painful, sometimes fatal

On 18 March 1965
Soviet cosmonaut
Alexei Leonov made
the first space walk
from the Voskhod 2
spacecraft

problem that afflicts divers who surface too quickly. So the space-suit pressure was set for a compromise, 6 lb per square inch. Leonov, at this pressure, pulled himself through the hatch and floated away, the first man ever to be free in space, and absorbed by the joy of it. After a few minutes he turned, grabbed a handrail near the hatch and tried to poke his feet back in. But, in his pressurised space suit he couldn't bend enough to do it. He almost established another Soviet first – as a man overboard in space.

There was only one solution. He lowered the pressure still further to 4 lb per square inch, when he could bend just enough to fight his way in. He arrived with his visor fogged and his space suit soaked in sweat, but this didn't feature in his report. He merely described an exhilarating space walk that had gone as rehearsed and expected.

Leonov's troubles weren't over. Soviet spacecraft are usually brought down by retro-rockets that are fired automatically, but a day later, when Voskhod 2 was due to land, the automatic system failed. On the next lap, the second astronaut, Pavel Belyaev, fired the retro-rocket himself, but being a lap late, the spacecraft was off-course and landed 2000 miles

54

away from the intended place – in deep snow in the Ural mountains. The crew had to survive a cold night, surrounded by wolves, before the rescue team found them.

So, while the Americans were developing new spacecraft for new tasks, the Russians, desperate to avoid being overtaken, had been riskily stretching the capabilities of the old craft. Though the world knew none of this at the time, the Russian space scientists recognised that their manned flight programme had reached a plateau. There were no repeats of the Voskhod 'triumphs' – no more Voskhod flights at all. The next flight, in Soyuz 1, was in 1967, a couple of years later. And that flight was a fatal disaster.

The American approach to a moon landing was cautious and systematic to an extreme, but their steady development of the Gemini series was a success that gave them the technological experience they needed. There were two launches of an unmanned Gemini, on 8 April 1964 and 19 January 1965, before the first manned flight on 23 March 1965. Even this was modest. Virgil Grissom and John Young made just three orbits in 4 hours and 53 minutes, but the trip was a success and even notched up a first – it was the first spacecraft in which the astronauts could switch orbits. While the solo passengers in the Mercury series had had virtually no control over their craft, the voyagers in the Gemini craft were developing the control that would be needed for a trip to the moon.

In the second Gemini flight – called Gemini 4, as the unmanned tests were counted – launched on 3 June 1965, the Americans James McDivitt and Ed White attempted to extend this manoeuvrability to rendezvous with another object in space, the second stage of the Titan rocket that had launched them (the second stage accompanies the spacecraft into orbit). Unfortunately the plan failed, and the spacecraft wasted a lot of fuel chasing the rocket.

However, the planned space walk was a success. On the third orbit Ed White, dressed in a space suit, tethered to his spacecraft, let the oxygen out of the spacecraft, opened the hatch and slid out. He propelled himself around for a few minutes with a hand-held jet-gun, and when the fuel ran out he experimented to see what it was like to move about in space. After a total of twenty-one minutes outside, Ed rejoined the craft. It eventually splashed down on 7 June after a total of sixty-four circuits.

Gemini 5 had a frustrating flight. Gordon Cooper and Pete Conrad took off neatly enough on 21 August 1965 with the intention of demonstrating a rendezvous technique: they were also to test a new kind of electric battery – the fuel cell – in space. This proved to be their undoing. The fuel cells malfunctioned so badly that the rendezvous was abandoned. The fuel cell did eventually work properly, but by then there were faults in the manoeuvring jets. NASA organised a rendezvous with

*Edward H. White
performing his
spectacular space feat
during the third orbit
of the Gemini 4-Titan
flight*

an imaginary target and claimed to be totally satisfied with the result, but no one else was.

But the flight had one success: it confirmed the quite remarkable ability that astronauts had to see objects on earth. From the earlier flights astronauts had reported seeing smoke from chimneys and the wakes of ships, yet these should have been too small to be seen from 100 miles up and more. Gemini 5 had cameras that showed striking detail but, more convincingly, the astronauts were able to pick out the launching and flights of Minuteman missiles on test. This was detail which it was theoretically impossible to see from space: the fact that it was visible encouraged those who were interested in surveillance from space satellites.

Gemini 6 was to be yet another attempt at rendezvous. An Atlas rocket with an Agena spacecraft as a target was launched on 25 October 1965, and the manned spacecraft with Walter Schirra and Tom Stafford aboard was to follow. Unfortunately the Agena didn't go into orbit, so yet another rendezvous attempt failed. The Gemini 6 wasn't launched.

56

Photograph of Gemini 7 spacecraft taken through the hatch window of Gemini 6 during rendezvous manoeuvres after take-off on 15 December 1965

A plan for a second launch was made soon afterwards because the Americans wanted to keep to a schedule of progress, despite the run of failures at rendezvous. Gemini 7, which was launched on 4 December 1965 with Frank Borman and James Lovell aboard, was intended as a test of man's ability to stay in space – it would remain in orbit for a fortnight. But it could also be used to test the rendezvous technique. Gemini 6 was still waiting for take-off.

57

The first attempted launch had to be cancelled, but on 15 December Gemini 6 took off, went into the same orbit as Gemini 7 and set off to meet it. This rendezvous was a success. There were no plans to 'dock' with one another – to link up in space – but the spacecraft did everything but this. It was manoeuvred to within a few feet of Gemini 7 before returning to earth, to splashdown on 16 December. Gemini 7 went on to establish a new space-endurance record with a trip of more than 330 hours. The Americans were beginning to get it right.

Gemini 8, launched on 16 March 1966 with Neil Armstrong and David Scott, looked like another successful step at first. Six hours after take-off it achieved the first spacecraft docking in history when it linked neatly and smoothly with an Agena spacecraft that had been separately launched. Then trouble started. The plan was to stay in orbit and make another space walk, but the coupled craft suddenly started to zig-zag and turn end over end. One of the Gemini control jets was failing to function. With difficulty Neil Armstrong separated his craft from the Agena and shut the jet off; by then, most of the fuel had been wasted, so the trip was cut short and the craft splashed down after a total of less than eleven hours.

Though the Americans were making progress, it wasn't smooth. Gemini 9, launched on 3 June with Tom Stafford and Eugene Cernan, was intended to show how easy rendezvous and docking were by making three encounters with an Agena target. Unfortunately the target's docking mechanism jammed and the astronauts had to be content with close approaches. This flight was also designed to extend the new skills in spacewalking. Cernan went out on the second day with an elaborate personal jet system on his back, and duly spent more than two hours manoeuvring in space. Some of the equipment he had intended to work on in space had proved tricky, so Cernan was tired by the time he returned to the spacecraft. Then there was near disaster. He found getting back into the craft surprisingly difficult, because his space suit was inflexible in the vacuum of space, and he arrived inside exhausted, his visor clouded with mist.

The Gemini 10 trip, with John Young and Michael Collins, was reassuring. It took off on 18 July and docked with an Agena. The craft separated, the Gemini went off to rendezvous with another Agena, and Michael Collins scored another first by spacewalking to the Agena, returning with an instrument that had been attached to it. Gemini 11 again achieved a successful docking, but also demonstrated further the problems of spacewalking. Richard Gordon had to cut short his trip because of exhaustion.

The last trip in the series was a success. James Lovell and Edward Aldrin were launched in Gemini 12 on 11 November 1966 and successfully docked with an Agena. More important, Aldrin worked out a way of moving about on the spacecraft – making the maximum use of hand-holds, anchoring himself with two tethers so that he could work. He tried out a series of tools on a metal panel attached to the front of the spacecraft and on another fitted inside the docking adaptor, and he conserved his energy so effectively by taking rests that he had no problems either with his tasks or his return to the Gemini.

The Americans were now ready for the next series of steps to the moon, and confident. So, too, as far as the West could tell, were the Russians. Tragically, both nations experienced disastrous, fatal failures.

Gemini 9, launched 3 June 1966

4

DEATH AMONG
THE ASTRONAUTS

By a macabre coincidence, there were fatal space disasters in both America and Russia at much the same time. They were completely unexpected. Space travel, in terms of thousands of miles travelled per death, is the safest form of transport known, though this is perhaps not too surprising as the enterprise has attracted the best scientists, the best engineers and the best designers of the period.

At times, however, caution has been sacrificed for progress. Space enthusiasts have been anxious for achievements, and tempted into asking astronauts to try a slightly more experimental project than was wise. And politicians needing a space triumph at a particular time could be tempted to press space engineers and planners to hurry with a project. Such an approach risks disaster.

For a long while, the Russian habit of reporting only successful achievements led to persistent rumours of just such disasters, of secret launchings and of astronauts lost in orbit. It is now virtually certain that these rumours were ill-founded. Rocket launches are easily recognised by the spy satellites that survey the Soviet Union; the orbiting spacecraft have been monitored by all sorts of groups, amateur and professional, and none of these noted any secret trips. Furthermore, it is likely that, by now, some defector would have produced a convincing report about Russian space failures had there been any substance in the rumours.

What a spy satellite might not have spotted was a Soviet launch-pad disaster, and there is fairly conclusive evidence that a politically motivated launch led to just such a disaster in October 1960. A 'launch window' for the planet Mars opened that month. Launch windows are periods when launches are particularly attractive, usually because they are the times when least fuel will be required for a mission. Launch windows for Mars occur only every two and half years, so there would be a long wait for another October window. As October is a historically significant month in the USSR, several space launches, including the world's first had been made at that time of year. In September 1960 rumours of an unmanned space probe to Mars were buzzing around Moscow. The Russians were particularly anxious for success because two previous Mars probes had apparently failed.

At the time, Khrushchev was attending a UN meeting in New York – he arrived on 19 September – and a sailor who defected from his ship said that his baggage included models of a spaceship. Khrushchev, of course, liked to encourage space triumphs that could be used as propaganda and, if he was expecting one, there is every likelihood that he would have equipped himself with models of the appropriate spacecraft. Also, Russian space-tracking ships had been deployed – another fragment of evidence to support the rumours of the Mars trip.

But no launch was reported: Khrushchev returned to Russia in mid-October and so did the tracking ships. It seemed as if a launch had been cancelled. Then further rumours started – this time about a disaster. Such rumours were not exactly rare, but years later there was sufficient evidence to suggest that they may have been true. When the Russian biochemist, Zhores Medvedev, defected to the West in 1976 he described a number of unreported Soviet disasters, including a nuclear waste explosion in the Urals that we now know did happen. Another of his reported disasters was a launch-pad explosion that had destroyed a Mars probe. According to Medvedev, the designer Sergei Korolyev was under heavy pressure from Khrushchev to achieve this launch before the window closed for a decade. Some of the best Soviet rocketeers were said to be involved, including the missile expert Field-Marshal Mitrofan Nedelin. As the countdown neared zero, Korolyev and Nedelin were in the security of the control bunker, but at zero the rocket didn't ignite. If the rumours and Medvedev's account are true, Nedelin's anxiety led him to leave the bunker with a team of technicians to inspect the rocket. They had waited a while before doing this, but not long enough: the rocket belatedly exploded, killing Nedelin and his team. This cannot be proved, but according to official records Nedelin did die, allegedly in an air accident on 24 October 1960.

There was no secret, and no doubt, about the first and only fatal US spacecraft disaster. In this case the spacecraft wasn't even about to be launched. It was undergoing its preliminary check. The spacecraft in question was the first Apollo, the date 27 January 1967. The crew – Gus Grissom, Ed White and Roger Chaffee – were aboard, working through tests of the equipment. They were breathing pure oxygen, which had been chosen as an alternative to air because it simplified the machinery of what was called, ironically as it turned out, the life-support system. There were gauges outside that measured the rate at which the oxygen was being used. At 6.30 p.m., the rate rose quickly – very quickly indeed. Ed White's heartbeat, which was also checked from outside, speeded up as if he were frightened or at least excited. And at 6.31 there was a yell from inside: 'Fire in the spacecraft,' followed by 'Get us out of here.'

Those outside saw a burst of flame, and hands reaching for the bolts that held the escape-hatch shut. As we know, the earlier quick-release system had been abandoned after it had malfunctioned – with, ironically, Gus Grissom aboard – and allowed the spacecraft to fill with sea water. Now no one inside could open the hatch, and the spacecraft was so hot that no one outside could get in for six minutes. All they found then were the charred remains of the three astronauts.

The report on the accident was damning. The immediate cause of the fire had been an electrical spark; the reason it had spread so quickly was the atmosphere of oxygen inside the spacecraft, and a quite remarkable amount of plastic substances that would have been flammable even in air. The investigation showed that however firmly the planners at the top may have demanded total safety and immaculate workmanship to achieve it, further down the line an almost reckless sloppiness had crept in.

The fire started below Gus Grissom's couch; the spark almost certainly originated from some wire where the insulation had been eroded. No one is certain about which section of the wiring was involved, but it was probably a bunch that went under a rather close-fitting door. It is difficult to believe that the space industry could work in a way that every do-it-yourself home electrician knows is dangerous but, however sophisticated a system may be, there is always the danger of human error.

Two particular pieces of plastic equipment were involved, both put into the spacecraft to reduce the practical difficulties of living in the weightlessness of space. The craft was lined with a fine nylon netting to keep small objects from floating into corners and crannies, and there was Velcro strip to hold small bits of equipment in position. These materials both burned fiercely.

The heat warmed the oxygen inside the craft and drove the pressure upwards so rapidly that the spacecraft split, all within fifteen seconds of the start of the blaze. The explosion produced waves of hot gas that set alight the rest of the plastic inside, until eventually the flames needed oxygen more quickly than it was being supplied. Now the flames extinguished themselves, filling the cabin with carbon monoxide and the other lethal gases that come from burning plastics. It was these gases, almost certainly, that killed the astronauts, although this was only a matter of timing. They would never have survived the effects of their burns.

Had the spacecraft been filled with air, rather than oxygen, there would have been no fire, certainly no holocaust. NASA obviously needed to establish how a minor routine test, where no danger at all was anticipated, could end so tragically.

The list of simple avoidable mistakes was enormous. The hatch that sealed the spacecraft opened inwards, for example. This was logical – it meant that the pressure inside the cabin sealed the hatch more securely while the craft was in the vacuum of space. But it also meant that the hatch was held shut by the pressure generated by the fire. Even in ideal conditions, it took a minute and a half to open the hatch.

At the time of the fire the Apollo capsule contained more than 70 lb of plastic: the official enquiry proved that quite a lot of this hadn't been properly tested to see if it was fire-resistant. Ten pounds or so of flammable material was junk, padding for the hatch to rest on during maintenance – cotton rags and waste. When plastic materials are used in a spacecraft, safety regulations determine that they should be confined to areas isolated by firebreaks – and, of course, there should not have been any flammable waste in the spacecraft when the astronauts sealed the hatch.

NASA's investigating board found many examples of poor workmanship in the wiring of the spacecraft. It had been forced through narrow gaps with frequent sharp bends, and sometimes jammed up against panels. In either of these places the insulation could have been worn away by vibration, but in some cases there was hardly any need of vibration to damage some of the wiring: it was laid across the floor, to be stepped on by workmen and astronauts.

Then the investigation revealed that the cooling system had had a long history of failures. The liquid that circulated in it, similar to the liquid in a domestic fridge, was flammable and there had been frequent leaks. This, like the other faults, could well have put an actual flight at risk, and it was yet another fault that would have been spotted by a competent system of inspection and management. In addition, it was pointed out that there was no quick escape route for the astronauts, and insufficient fire and medical equipment on the launch pad.

In one way the enquiry was inconclusive. No one knows exactly why the work was so badly done. The Apollo project was working against time – it had to meet the date set by President Kennedy – but there was no great pressure at that point. In any case, the need for speed could never justify sloppiness. There were suggestions that the makers of the command module, North American, didn't really have a good enough track record to justify their getting the contract, but NASA seemed more anxious to get on with the job than to find out who was to blame, because a widely publicised post-mortem might have disturbed public confidence in the whole Apollo project.

NASA now had to go back to the drawing-board and redesign the Apollo command module; the Review Board, set up after the accident, had proposed some 5000 alterations. It cost NASA a stunning $40

63

million to redesign the hatch alone; they also replanned the wiring and systematised the use of plastics. The result was that there were no manned launches for more than a year and a half. The next manned flight was Apollo 7 – the earlier Apollo flights had been unmanned – which took Walt Schirra, Donn Eisels and Walter Cunningham into a 260-hour space trip on 11 October 1968.

The first recorded Soviet space death, also involving a new craft, the Soyuz, occurred within a few months of the fire in Apollo. Ironically, the Soviet scientists were confident of success on this particular flight and for once described in advance what they intended to do. There would be two spacecraft in orbit, they would rendezvous and link, and the crew would swap spacecraft. This would be a vital forerunner of an attempt to land on the moon, and a striking achievement in its own right.

Soviet astronaut Vladimir Komarov, pilot of the new Soyuz 1 spacecraft launched on 23 April 1967, which crashed to earth killing Komarov

*Launching of a Soyuz
spaceship*

65

On 23 April 1967, Soyuz 1, with Vladimir Komarov as pilot, took off. He was a relatively experienced astronaut, having been commander of the first, and only, three-man Voskhod 1, and the first Soviet astronaut to make a second flight.

There was certainly no chance of his making a third. After only a day, the Soviet news agencies announced that the flight was to end. For twelve hours after that there was no announcement. Then the world was told, 'Komarov is dead.' The parachute lines on his spacecraft had tangled and the spacecraft had crashed to earth. The Russians were laconic. They didn't say why the lines had tangled, but there was obviously some significant problem that took a long while to solve.

By combining not particularly reliable information from intelligence sources with what engineers know about the Soviet space techniques, it is possible to suggest a sequence of accidents. The Soyuz power supply – solar cells – failed in part. Komarov as a result missed his scheduled re-entry, went around the earth again, missed again and set off on a third lap. He then came down rather steeply and set the spacecraft spinning to keep it on its path, in the same way as spinning a rifle bullet keeps it in accurate flight. There is no great difficulty in using the thruster jets of a spacecraft to start it spinning, and equally no problem in using the same jets to stop the spin before the parachute is due to open. But Komarov would have been spinning as he hit the atmosphere, and the combination could have caused a temporary and disastrous blackout. We don't know, may never know, what happened, but the theory that he landed too steeply is supported by the fact that the craft landed 600 miles short of its target. It took the rescue crews a long while to find the smashed craft and its dead pilot.

5

RACE FOR THE MOON

The two disasters had to be accepted and understood before the race to the moon could start again. Even though the public accepted that space exploration was dangerous – and astronauts were seen as heroes – a repetition of the unforgivable errors that killed the three men in Apollo 1 would be intolerable to public opinion, and to NASA and the American government that backed it.

And though Russian public opinion is never really heard, it clearly echoed the American. Russian leaders also appreciated that there was no national prestige to be won from a foolish disaster in the space contest. The exact effect of Komarov's death on Russian planning is unknown. The one ascertainable fact is that all manned space flights were stopped for eighteen months – Soyuz 2 was not launched until 26 October 1968.

Morale at NASA and in the rest of the USA needed a boost, and the most pressing priority in the race to the moon was the development of a new launcher. Here there was a real prospect of achievement as an enormous new rocket, Saturn V, was being designed by Wernher von Braun. The first stage – the lowest as it sat on the launch pad – was to be 138 feet high, 33 across. It was a cluster of five engines, burning kerosene (paraffin) and oxygen to generate a thrust of seven and a half million pounds. This stage was to take the assembly a modest 37 miles upwards, when it would be discarded.

The design of a large liquid-fuelled rocket is considerably more complex than that of a small one; the particular problem is the vibration caused by the burning of the rocket fuel. The burning gases set up waves of pressure, and these can produce explosions. The work on Saturn V involved a lot of trial and error and a fair amount of inspired guesswork. Eventually, though, the design was complete and the first stage burnt successfully in more than two thousand trials.

The second stage was more modest in scale – 81 feet high and still 33 feet across – but more daring in design. Like the first stage it had five engines, but they burned hydrogen and oxygen, both carried as liquids.

It would take the spacecraft to 100 miles above the earth before being discarded, when the third stage would take up the burden. This was a single-rocket engine, 58 feet high and 22 across, also burning oxygen and hydrogen, and it was designed to take the spacecraft into orbit around the earth. The rocket would then shut down – no energy is needed to keep a spacecraft in orbit – but, when the time came to blast the craft out of the earth's orbit on its way to the moon, this third stage would have to fire again. It is difficult for a liquid-fuelled rocket to fire while it is in orbit because the liquid fuels are drifting weightlessly inside the rocket; the third-stage rocket used the explosion of four solid-fuelled rockets to drive the liquid fuels towards the pumps that would deliver them to the burners.

The Saturn V engines were designed and tested with quite remarkable thoroughness. NASA knew that when its Saturn V came to be tested, the public would base its judgement of the whole space effort on the results; equally, NASA felt confident that its careful testing of every individual piece of the rocket would guarantee the success of the whole rocket.

The first Saturn V flight – it was labelled Apollo 4 – took off on 9 November 1967. The countdown had run perfectly and the assembly took off at 7 a.m. The three stages burnt correctly and the third stage with the spacecraft went exactly as planned, into orbit 115 miles above the earth.

Then NASA showed its courage. It had already successfully fired the most powerful rocket in the world. Now it set out to test the ability of the third stage to set off to the moon, and the ability of the command module to withstand the heat and shock of re-entry. The third-stage rocket was fired again by remote control, and it took the spacecraft into an orbit more than 10,000 miles out. The spacecraft's own engines, the service module, increased the diameter of the orbit by another couple of thousand miles and then sent it back towards the earth, eventually driving it to 23,000 mph. These manoeuvres all had to be successful in order to prepare the way for an eventual moon landing and return; indeed, they had to be successful on this first launch for NASA to regain its credibility and maintain its financial backing.

Everything worked perfectly. The command module dropped into the Pacific Ocean 8 hours, 37 minutes and 8 seconds after launch. It was, in NASA's jargon, 'all systems go for the moon'.

They had first to overcome what scientists call a non-trivial problem about landing on the moon. Getting there was straightforward, at least

NASA's first Saturn 1B rocket successfully launched an unmanned Apollo spacecraft on 26 February 1966

in theory – as long as there was a properly designed life-support system, a powerful enough rocket and adequate computing facilities. There were no new principles involved and everyone realised that, given enough time and money, man could certainly *get* there. But no one knew if he could *land* there.

The fact was that they didn't know exactly what covered the surface of the moon. The worst possibility was dust. The surface of the moon had been subjected to extremes of temperature and bombarded by meteorites and cosmic rays throughout its history; it had no atmosphere to protect it. This had certainly produced some dust, but there might be a layer 20 or 30 feet thick. When the moon lander touched down, it might simply sink beneath the dust without trace: the first moon landers might disappear for ever in the quicksands of the moon.

Preliminary surveys of the moon had started much earlier, using unmanned spacecraft for the necessary reconnaissances. They were cheaper, simpler and lighter than the corresponding manned ones, and they could establish what sort of manned spacecraft should be sent to the moon.

The history of unmanned spacecraft had been peppered with failures. The earliest were launched by Russia but, as Russia rarely announced what a spacecraft was intended to do, there is no way of being certain that Luna 1, which took off on 2 January 1959 and missed the moon by 3000 miles, was intended to do anything else. Luna 2, which was launched on 12 September of the same year, crashed into the moon and, again, may have been intended to. They were both simple spacecraft with no cameras and few instruments.

Luna 3, launched on 4 October 1959 – two years after the first Sputnik – was quite different. Carrying a camera, it had looped around the back of the moon and taken photographs of the side never seen before, processed the films and sent the pictures back to earth. It was a stunning achievement. The rear of the moon, it turned out, was much smoother than the side we see – few craters or mountains, and none of the great maria (seas).

America, after three failures in the Pioneer series, had sent Pioneer 4 past the moon and on into space, but this was no great achievement. Then, in 1961, the USA had initiated the Ranger series – craft that were intended to take pictures as they went towards an eventual crash on the moon. The first six were uniformly disastrous: they either failed to take off, or the electrical systems burned out when the cameras came to be used. This run of failures was very dispiriting for the Americans, and there was talk about shutting down the whole vastly expensive project.

Ranger 7 changed the whole outlook. It took off on 28 July 1964 and duly raced the quarter of a million miles to the moon. In order to get pictures of the moon in ever increasing detail as well as pictures showing large areas, the TV cameras were switched on when the craft was about a thousand miles from the moon and left running until it crashed on the moon's surface. The final picture was transmitted from a height of about a mile. A total of more than 4300 pictures was sent back: they showed the moon, or at least that part of it around Ranger 7's target, in marvellous detail – the equivalent of seeing it from half a mile up. There was a mass of craters, some only three feet across, and it was also possible to identify the detail of the rays – the long whitish streaks that seemed to radiate from the craters. There were some sharp-edged craters, presumably produced by meteorites; others were softer and might be surrounded by dust. The so-called sea – the Mare Nubium – where the space probe crashed, seemed to be smooth, though not necessarily rigid.

Ranger 7 didn't answer all that many questions about landing on the moon, but it showed how they could be answered. What was needed was a series of similar flights that could map some areas and test the terrain, and a series of orbiting probes that would produce a complete map of the moon's surface. Even mapping from above gave hints: Ranger 7's observations suggested that the causes of the maria and the craters on the moon were meteorites, so the shape of the craters made it likely that the surface was rigid. Ranger 7 was a success and, in terms of the quality of pictures from the moon, a first.

On 17 February 1965 Ranger 8 set off, directed at the Sea of Tranquillity. It sent back more than 7000 pictures of an amazing detail – objects only five feet across were discernible. The pictures showed mountains – many of them twice as high as Everest – around the seas, and also what were clearly holes made by meteorites; these sometimes seemed to have been altered in shape by a later impact, producing a second hole. More significantly, there were also patterns showing that molten rock had travelled over the surface. The moon might be stable now, but there had been volcanic activity in the past.

There were equally marvellous pictures from Ranger 9, which was launched on 21 March 1965 and crashed, three days later, into the crater Alphonsus; these pictures were transmitted worldwide on television as they arrived. However, by now close-up pictures of the moon were becoming commonplace and, although they were fascinating and gave some idea of the geological history of the moon, they still did not answer the one question vital to a lunar expedition: was the moon's surface strong enough to bear the weight of a spacecraft and its astronauts?

The Russians answered this question with a marvellously simple experiment carried out on a dramatically sophisticated space probe. They had been trying to achieve a soft landing for some years, but Lunas 4, 5, 6, 7 and 8 had all failed in different ways. Luna 9, which was launched on 31 January 1966, was directed towards the Ocean of Storm, and for most of its trip it travelled like any other probe. Then, 50 miles above the moon's surface, it fired a retro-rocket. This slowed it down and, when it was only a few feet above the surface, the satellite dropped a large, two-hundredweight iron ball. Then the probe settled gently on to the moon, lowered a TV camera and sent back pictures, for the first time ever, from the surface of the moon. The Russians had achieved the first soft landing.

Luna 9 didn't sink into the moon's surface, so it appeared that at least some parts of the moon weren't feet deep in dust. And among the TV pictures sent back was one of the two-hundredweight ball which had hit the moon and rolled along. It would, the Americans were delighted to learn, be possible to land on the moon and walk on it.

Luna 9's landing also depressed the Americans as it seemed to indicate that the Russians were interested in a moon landing; it also showed that in at least some of their technology they were ahead. The Americans had plans for a soft landing – the Surveyor series – and had been struggling to achieve one for nearly a couple of years. So far, every launch had failed.

The American problem was one that is common to space exploration: the probe that the scientists wanted to launch needed a more powerful rocket than the engineers knew how to build at the time – von Braun's Saturn was still unfinished. Eventually the load demanded and the power available coincided, and Surveyor 1 made a soft landing on 2 June 1966. The pictures produced were much better than those of the Russians, and the information sent back was marvellously detailed and valuable. Surveyor's 'footpads' were designed so that they would put the same pressure on the moon's surface as a spaceship. Ingeniously, the TV camera was designed in such a way that it could photograph the pad and the area around it, and it recorded that the pads made hardly a mark on the surface.

The rockets available for launching Surveyors gradually became more powerful, making it possible for more instruments to be taken up. These included, for example, mechanical scoops that picked up lunar soil, kneaded it and then allowed the ball thus formed to break apart. It turned out that lunar soil had much the same consistency as good, damp terrestrial earth.

The Russian equivalent of the Surveyor's 'soil grabber' was carried on Luna 13. This landed on 24 December 1966 and was fitted with a drill that was driven into the ground by a charge of explosive: the craft then transmitted information about the quality of the ground under the drill. It reported that the first foot or so of moon soil had about the same physical properties as an average sample of terrestrial soil, exactly what

opposite, The Soviet Luna 9 automatic station

any future moon landers wanted to hear. These results were confirmed by Surveyor pictures which showed the rocks scattered around craters and could estimate the distance the bedrock was below the visible surface: it was never far. And, while the Surveyor pictures frequently showed rubble, there was never dust.

The last three of the five Surveyors brought off a thoroughly distinguished scientific first. A rather ingenious arm lowered a source of radioactivity – alpha particles – so that the rock was irradiated. This made it temporarily radioactive and, by investigating the rays the rock sent off, scientists on earth could determine the chemical elements in the moon rock. They found that the rock was basalt, common on earth and produced here by the heat of volcanic action. It might have been formed on the moon by the heat that meteorites would produce when they hit its surface, but there was a general feeling that there had been volcanoes on the moon.

As well as making detailed pictures of small areas of the moon, and getting a reasonable knowledge of its surface, NASA scientists needed probes that would orbit the moon. A moon landing would be made from orbit, so it was reasonable to see if there were any problems about the manoeuvre. Orbiting probes could check for dangerous amounts of radioactivity, and also map the moon and scan it for landing sites. The American programme of orbiting probes overlapped the Surveyor series.

The Russians had long before put a lunar probe into orbit for the first time. Luna 10 was launched on 31 March 1966 and went into orbit three days later. It discovered that the moon had a weak magnetic field and was surrounded by an extremely weak belt of radiation.

The Americans, typically, were second but more successful. The five flights in the Lunar Orbiter series started with a launch on 10 August 1966. The Orbiters were elaborate spacecraft – they carried a number of TV cameras, arranged to give overlapping pictures that could be put together to give a three-dimensional effect. They mapped the moon in extraordinary detail – showing craters and boulders, the tracks left where boulders had rolled down an incline, and winding channels along which liquid lava had flowed.

The Orbiters also confirmed the Russian findings about the other side of the moon, that it wasn't simply a continuation of the side we can see: the surface was much simpler. The significance of the fact that the far side had none of the great maria was not clear. After all, the only difference between 'our' side of the moon and the other, so far as position in space goes, is just that: one side faces the earth, the other doesn't. It is imaginable that the earth shields our side from meteorites, but that would produce the opposite effect. The only likely cause of any difference between the sides was the effect of the earth's gravity, or perhaps some minor difference due to a shading from some radiation. Neither of these gave a good explanation of the visible difference. As so frequently happens, research had produced a further problem rather than solving one that already existed.

The Orbiters discovered another group of mysterious features. As they orbited the moon they were monitored very closely and it turned out that they wobbled slightly as they went. They were being kept in orbit by the moon's gravity, and there seemed to be a dozen patches where this was slightly stronger than expected, probably due to patches of heavy material near the surface. These concentrations of mass, called mascons by NASA, were a complete mystery, though there was a reasonable possibility that they were enormous meteorites buried under the surface. What was important about them at the time was that they would clearly upset the track of any manned spacecraft going round the moon. Their effect would have to be taken into account in detailed navigational calculations for the command module and the lunar bug.

Though there were still mysteries about the moon, there seemed to be no serious problems impeding a landing. It had been established that spacecraft could land there and astronauts could walk on its surface. And, from the detailed maps that the Orbiters produced, NASA was able to select a landing site: the Sea of Tranquillity.

While every part of space exploration poses problems, the design of the moon bug – officially the lunar module – had special difficulties of its own. It had to be capable of delicately controlled ascent and descent, and of manoeuvre and rendezvous; it had to support two exploring astronauts for at least a couple of days; and yet it had to be light enough to be carried by a Saturn V. The Saturn could support a load of 14 tons, but it was difficult to restrict the lunar module to this weight. It could be saved by making the module structurally very weak – it would not, after all, have to stand on its own four spidery legs until it was in the weak gravity of the moon – but it therefore couldn't be tested on earth.

Its first flight was called Apollo 5 and took place on 22 January 1968. The bug was unmanned and its two engines were tested by remote control. The descent engine worked perfectly even though it was, for the first time in history, a controllable rocket engine. This was a necessity: the astronaut had to be able to manoeuvre with the delicacy of a helicopter pilot as he landed. For all its novelty the descent engine worked reliably, but the ascent engine was endlessly troublesome. The only time when moon landers have to rely on one piece of equipment, with no stand-by and no chance of rescue, is when they take off from the moon. A failure at this point would leave the astronauts forlornly shipwrecked. NASA eventually had to replace the troublesome engines with totally new ones, when the lunar module behaved perfectly under test. Another of the steps towards the moon had been taken.

Now the Saturn V had worked, the command and service modules had worked, and the lunar module had worked, all unmanned. The prospect was one of plain sailing. As a piece of scientific caution, though, NASA ran another test launching of the Saturn V on 4 April 1968: this was labelled Apollo 6. The results were unhappy: the second stage stopped firing prematurely and so did the third stage. This third stage, that would eventually have to be refired to take the spacecraft out of orbit and off towards the moon, would not restart. Fortunately, a rocket engine is extremely well equipped with instruments, and every one of those instruments sends its reading back to earth where it can be captured in print and analysed by computer. The NASA engineers quickly discovered that the engine failures were caused by simple, easily corrected faults – one a wiring error, one a fuel leak. The test was counted as a success and it was decided that there would be a crew aboard the next flight of a Saturn V. However, before that could happen, the command and service modules had to be tested with a crew aboard. As this test wouldn't go outside an earth orbit, it didn't need a Saturn V: the less powerful Saturn I would be adequate. The crew was Walt Schirra, Donn Eisels and Walter Cunningham, and their trip test, Apollo 7, took off on 11 October 1968.

It was a perfect flight. The spacecraft duly went into orbit and separated from its Saturn rocket. It then showed its manoeuvrability by approaching and moving away from the rocket, using the service module's engines for power. The crew made observations of the ground and of the cloud formations of a storm: they even, for the first time, had a hot drink – coffee – in space. The only problem was that all the astronauts developed colds, which are worrying in space because they might cause blockages in the tiny tubes from the ear that allow the pressure to be balanced as the astronauts descend from the low pressure in the craft to the normal atmospheric pressure of the earth. While the worry was justified, the astronauts landed without any ill effects. Now the rockets and spacecraft had all worked while unmanned, and the command and service modules had successfully carried a crew. All that was left to test, manned, was the lunar module and the Saturn V rocket.

FACING PAGE
above, *Soviet cosmonauts Yuri Gagarin and Valentina Tereshkova*

below, *The Russian Vostok-1 spacecraft*

For neatness and economy, both tests should have been made on the same trip, working for safety in a modest earth orbit. But there was a problem. The Saturn V could be tested without the bug on the scheduled date, 1 December 1968, but that trip would be almost identical to the previous Apollo 6 trip, only using a different rocket. So NASA became daring. Apollo 8 would take men round the moon. No man had been there before.

The Saturn V had been tested around the moon, and the command and service modules had worked well in earth orbit, both unmanned and manned. Sending men around the moon ought to work, but there was the risk of the unknown. The command and service modules hadn't ever been that far from earth. The life-support system might fail, or the service module's rocket might not work properly at the range it would be operating at. Yet all the machinery had worked well, or had misbehaved in an understandable manner. It was a reasonable next step towards the landing. And the Russians had just reported recovering two unmanned spacecraft that had been round the moon. It could well be neck and neck to send the first men round.

On 21 December 1968 Frank Borman, Jim Lovell and William Anders were sent off to the moon. Twelve minutes after lift-off Apollo 8 was in orbit around the earth. Though NASA wanted the men to go round the moon, the trip had a number of decision points. If there were problems, the spacecraft need not be committed to the next stage. Earth orbit was one such point, but the entire assembly seemed fine and the crew re-lit the third-stage rocket to send the spacecraft out of orbit towards the moon. Soon the crew had gone further than anyone before, and they attempted to describe what they saw to those on earth. Five minutes out of orbit, the third stage was shut down and the command and service modules continued on their own.

There were still no technical problems, but Borman and Anders became space sick. This was worrying as it might make the men useless passengers for the rest of the journey, leaving only Jim Lovell to bring the craft home. It wasn't an unanticipated problem, however, and the spacecraft contained a pharmacopoeia of anti-sickness and anti-diarrhoea medicines. The two astronauts were diagnosed by the most remote doctor in history, took their medicine and, after a day or so, started to feel better. Space sickness has been a disturbing possibility for every space flight since.

The command and service modules passed the point where the gravity of the earth balanced that of the moon and started to accelerate. Sixty-one hours into the flight, on the morning of Christmas Eve, the astronauts fired a course-correcting rocket. They would now skim the moon, 70 miles up. As the craft swung behind the moon the service module's engine was fired. This took the men into an elliptical orbit: later, the same engine was fired again, delicately, to put the craft into a near-circular orbit. It was man's first ever personal survey of the moon.

The three astronauts were still orbiting the moon on Christmas Day, taking hundreds of photographs and making a miraculous, though slightly disappointing, TV broadcast to Earth. Early that morning the service module's rocket was fired again to take them out of orbit, directed

77

towards the earth and the first test at full speed – around 24,000 mph – of the heatshield. This, like every other part of the spacecraft on the trip, worked perfectly, and they splashed down and were taken to the USS *Yorktown* on 27 December 1968. The prospects for a moon landing were good.

Next, the troublesome lunar module had to be tested with a crew, first in orbit around the earth. This was planned for the end of February 1969, and then postponed because the astronauts had colds. Apollo 9 – command module, service module and lunar module – took off on 3 March carrying James McDivitt, David Scott and Russell Schweickart into an orbit 119 miles above the earth. By then the first two stages of the Saturn V had been discarded: the third stage continued and protected the lunar module.

After nearly three hours in orbit, the command and service modules separated from the third stage, turned around in space, and docked on to the lunar module with what had been the front end of the command

The US unmanned Apollo 9 lunar module which landed on the moon's surface

module. The lunar module was drawn out of its sheath, and the third stage was discarded. The assembly continued in orbit for a longish period, then the USA notched up another first. Space walkers, to date, had used long umbilical tubes connected to air supplies and a cooling system on the spacecraft. But men would have to walk free on the moon. So Russell Schweickart tested, very modestly, a self-contained spacesuit. He stood on a small platform on the outside of the craft and then worked his way, hand over hand, across its surface. He remained, wisely, attached by a lifeline. There were no problems.

The fifth day saw the real test. Scott fired the explosive that shattered the bolt holding the lunar module to the command module. Then the lunar module with McDivitt and Schweickart aboard moved away, first a few yards, then fifty miles. The astronauts tested the descent engine, they tested the vital, unique ascent engine: everything worked perfectly. They rejoined the command module and crawled back into it. The lunar module – expensive, elaborate, fragile – was thrown away. In astronautics it is customary to test one device to see if a similar one will work.

The rest was routine. The astronauts returned, and NASA had to decide whether or not any other testing was needed before a moon landing could be attempted. In the end, a reasonable caution prevailed: the lunar module would be tested above the moon. This test, Apollo 10, carried Tom Stafford, Eugene Cernan and John Young into a position of marvellous frustration. They took off on 18 May 1969 and flew to the moon. Stafford and Cernan climbed into the lunar module and hovered in it over the Sea of Tranquillity – the planned landing place. They descended to a position nine miles above the moon – less than twice the height of Everest. Then, their duties over, they dutifully rejoined Young and the three of them came home. Someone else would make the final step.

A moon landing was now within reach. Apollo 11 was launched on 16 July 1969 on its Saturn V rocket and reached the moon's orbit on schedule. The voyage, including the first-ever colour TV broadcast from space, seemed routine. It involved marvellous accuracy, and the almost miraculous organisation of the most complex technology man has ever undertaken. But the problems had been mastered. The Apollo 11 astronauts – Neil Armstrong, Buzz Aldrin and Mike Collins – fired the service module's rockets as they went round the back of the moon, and, exactly as expected, the assembly – command module, service module and lunar module – went into lunar orbit. This really was a reliable routine: every time a lunar orbit had been attempted, it had succeeded. No other form of transport has that record.

However, while Apollo 11 was behaving perfectly, there was a metaphorical cloud on the horizon. Three days before its lift-off, the Russians

Apollo 11, riding a pillar of flame, rises to clear its mobile launcher

had sent a spacecraft, Luna 15, to orbit the moon. It was unmanned, so there was no question of beating the Americans to a landing. But why was it launched? What were the Soviet plans? Could they do anything to tarnish the American success?

The most serious problem was that they might, even accidentally, get in the way. One might think that the moon is large enough, and the skies above it empty enough, to accommodate two satellites in orbit, but the Apollo team certainly didn't want to add even slightly to their worries. So, in a remarkably straightforward way, Frank Borman from NASA telephoned the Soviet Academy of Sciences to ask for details of Luna 15's orbit. And, in an equally straightforward, though hardly predictable way, he received an answer: the Academy's President, Matislav Keldysh, reported an elliptical orbit ranging from 30 to 110 miles from the moon, although he would not specify why Luna 15 had been launched. Alarmingly, a Yugoslav news agency suggested that Luna 15 would soft-land on the moon, scoop up some moon rock, and bring it back to earth. It wasn't the same as putting people on the moon, but the Russians could, and did, start suggesting that real space exploration used unmanned probes – the rest, they said, was display. Few people, at that moment, were convinced. Anyway now that the Americans knew its orbit, they could avoid Luna 15.

As the command module (Columbia), lunar module (Eagle) and service module orbited the moon, the astronauts described the view. Another rocket burn was needed to convert the elliptical orbit into a near-circular one, then the preparations for the landing could start. Armstrong and Aldrin, who would, in that order, walk on the moon, crept through the connecting tunnel into the lunar module to check its systems. Satisfied with what they found they crept back, and all three astronauts took a short sleep.

The circular orbits took a couple of hours each. During the tenth, Armstrong and Aldrin again crept into the lunar module, for the last time. On the thirteenth orbit, Columbia and Eagle separated. This needed a short rocket burn that moved the pair apart, as planned. Armstrong announced this to Houston: 'The Eagle', he reported, 'has wings.'

The separation was gentle – the craft moved apart at two and a half feet per second. The two craft then appeared round the front of the moon side by side – the burn had been made out of sight. Armstrong fired the lunar module's engine to slow it down so that it would drop nearer the moon. As the astronauts continued to check their instruments, and NASA at Houston checked their much larger set of readings, there seemed to be only one problem. NASA sometimes found it difficult to communicate with Eagle and had to use the orbiting command module, with Collins aboard, as a relay. Eagle was still orbiting, and could still, as NASA described it, 'abort' and return to Columbia. But everything looked fine. Armstrong slowed Eagle further and it started on its unique, twelve-minute trip to Tranquillity.

With four minutes to go, a problem arose. Information was pouring earthwards from the instruments in Eagle and the computer at Houston became overloaded. It couldn't process the information quickly enough so, when a warning light came on, no one knew what it represented. Mission Control courageously decided that it was an instrument failing, not a failure of the actual working of the spacecraft. Armstrong was told to go ahead and land, using the automatic landing system.

This nearly brought them to disaster. Some faulty navigation had taken Eagle four minutes away from its proper landing place. Armstrong looked out of the window and saw they were landing in a bumpy boulder field. If Eagle tried to land there there was a real risk that it would tip over and, if it did, it would certainly never leave the moon. Armstrong took over and landed manually and safely. 'Houston,' he radioed. 'Tranquillity base here. The Eagle has landed.'

Eagle, the Apollo 11 manned lunar module, after its landing on the moon's surface

NASA's unimaginative plan was that the astronauts, having landed, should sleep for a few hours. This was now recognised as absurd, so the astronauts had their first meal on the moon and helped each other to put on their very bulky spacesuits. Later, when the astronauts were on the moon, in a vacuum, the suits became so stiff that the men sometimes couldn't see what they were doing. When Armstrong pocketed his first sample of moon rock, Aldrin had to direct his hands.

Eating, getting into the spacesuits and setting up the spacecraft for a quick getaway in case it was needed, took time – nearly six and a quarter hours. Then Eagle's hatch was opened and Armstrong squeezed out, cautiously testing the ladder to the ground. Like the rest of the lunar module's structure, this hadn't been tested on earth. Armstrong moved slowly down the ladder, taking a TV camera from its niche on the way. As he stepped on to the moon, in full view of the millions watching on television around the world, he uttered those immortal words: 'One small step for man. One giant leap for mankind.' He loped around a bit, describing what he saw – including his own historic footprints. Then he remembered, or was reminded, to pick up the 'contingency sample' – a mineral collection that would mean that something was brought back from the moon, even if the astronauts had to leave in a hurry.

Aldrin then joined his companion on the moon. They set up a TV camera on a tripod so that their activities could be seen on earth, and they triumphantly mounted a rigid US flag on a pole, though NASA emphasised that this was not an attempt to claim the moon, or even Tranquillity base, for the USA. They collected more rocks and they went through a mobility evaluation – in other words, they tried out various ways of moving on the moon: walking, loping, jumping. Walking

83

seemed most effective. Then they had a phone call from the new President, Richard Nixon: duly they saluted him and the rest of the earthlings, and continued their work. They set up various experiments – a moon-quake detector, and a laser reflector that would let scientists on earth measure the distance to the moon within an accuracy of six inches. Then they deposited the rubbish from Eagle, including the sewage, on the moon, climbed back into the craft and went to sleep.

Twenty-two hours after landing on the moon, Eagle's ascent engine was fired. This had no back-up: it had to work and it did. Eagle was carried upwards to rendezvous with Columbia. They experienced slight difficulties docking – Eagle wobbled a bit as it approached – then Armstrong and Aldrin rejoined Mike Collins for an uneventful sixty-hour trip to splashdown.

Eagle on its return flight from the moon

The command module splashed down on 24 July in what NASA disarmingly describes as 'Stable Mode 2', meaning, upside down, but it was soon righted. The astronauts were immediately taken off to quarantine: nobody could be absolutely sure at that time that there was no life on the moon that might contaminate us on earth. They made a brief TV appearance as heroes, were again congratulated by President Nixon, and then retired to a quarantine lasting 18 days. Eventually they re-emerged to celebration banquets and triumphal tours. The United States had won the moon race and justified President Kennedy's forecast. The space enthusiasts at NASA were already preparing for more moon trips, and for voyages further out in space.

Apollo 11 astronaut Edwin Aldrin places the first US flag on the surface of the moon

85

6

UNLUCKY 13

The successful landing and return of Apollo 11 demonstrated that the Americans had mastered the technology of manned space flight: space travel, it seemed, was reliable and predictable, if somewhat expensive. Of course, there were always uncontrollable natural hazards. Just after Apollo 12 took off on 14 November 1969 – less than a minute after lift-off – it was struck by lightning, twice. One of the astronauts, Charles Conrad, was recognisably awed – he was heard to say 'Everything in the world just dropped out' – but the trip continued, carrying Conrad, with Alan Bean and Richard Gordon, to the moon. Conrad and Bean landed on the moon, as planned, on 18 November, and left a package of scientific instruments to make observations that were radioed back to earth. The spacecraft arrived back on earth, without incident, on 24 November: the stupendous technological expertise that had made space travel possible made accidents almost unimaginable.

Then came Apollo 13. The number thirteen is considered unlucky in the USA, as in the rest of Christendom, but scientists and technologists could hardly take account of this fact. This next Apollo mission took off, after a minor delay, on 11 April 1970; the astronauts were Jim Lovell, John Swigert and Fred Haise. They duly went into orbit around the earth, and then set off towards the moon. The trip followed the planned pattern: the command and service module reversed itself to remove the lunar module from the third stage of the launcher, and then the whole assembly – command, service and lunar modules – was ready to land men on the moon for the third time. Though there was an army of controllers, technicians, scientists and engineers at Houston, the flight was going so well that many of them simply let the printouts run from their computers for the sake of the record. There was so obviously nothing wrong that there was no point in checking.

For safety reasons, when a moon-mission spacecraft first sets off from its orbit around the earth, it is directed into a 'free-return trajectory', meaning that it will simply loop around the moon and return. Supposing there is a fault – with the power or navigation, for example – the craft returns and, providing the command module's modest emergency equipment has worked, the astronauts are safe. But no failures were reported from Apollo 13 during the first twenty-four hours, so on 12 April a small rocket burn directed it away from the free-return trajectory and towards its target, the Fra Mauro hills, on the moon.

The following day the astronauts took TV watchers on earth for a conducted tour of the spacecraft – Fred Haise took a camera and guided the viewers around the rather cramped area of the lunar module, and then handed over to the Commander of the craft, Jim Lovell, who showed viewers pictures of the pilot, John Swigert, sitting at his console of instruments. Soon after the end of the broadcast a yellow warning light flashed above Swigert's head. The service module was carrying two tanks of hydrogen as fuel, and the pressure in one of them was low.

The service module contained the fuel and a lot of the other supplies for the command module. The hydrogen tank – a sphere containing hydrogen at a very low temperature – supplied the gas to a fuel cell where it combined with oxygen, also carried in a sphere in the service module, and generated electricity. A useful by-product of this was water. The oxygen sphere supplied the command module's life-support system as well as the fuel cell.

Throughout the trip it had been difficult to get accurate readings of the pressure in either of the hydrogen tanks or the oxygen tanks, which were also duplicated for safety. Instrumentation in space was always a problem because substances out there do not behave as they do on earth. The difficulty in this case was that very cold gases and liquids don't exert their full pressure. To solve this problem, the gas containers were fitted with heating wires that could be switched on to give the gases what NASA called a 'cryogenic stir'. As one of the hydrogen tanks was reading low, they were both given a stir to check the reading. The NASA controller at Houston also decided that the second oxygen tank should have a stir as it had steadily been giving a nonsensical reading of more than full, and for the sake of completeness the other oxygen tank was included. Swigert was asked to stir them all. He closed four switches on his instrument console and switched on the four heaters.

The controller at Houston, Seymour Liebergot, who had initiated the cryogenic stir, examined the effect it was having on the hydrogen gauges. Because of the way the warning lights worked they could draw attention to only one problem at a time: nothing suggested that any other pressure gauge might show anything out of the ordinary, and Liebergot didn't look at the readings on the oxygen gauges. One of these, though, was rising quickly and dramatically.

A few seconds later, Swigert reported to Houston, 'We seem to have a problem.' He had felt a strong, unusual vibration. Haise, who had been in the tunnel from the lunar module, had been shaken by an unusual movement, and Lovell, the Commander, had actually heard a loud bang. It was still 13 April, just after 9 p.m. Central Standard Time.

Though the astronauts knew that something had gone wrong – there was by now an amber warning light indicating an electrical fault, as well as the low-hydrogen light – no one at Houston shared their certainty: in fact Houston was enormously biased against accepting that fact. This was, after all, the thirteenth Apollo trip, and that series had been initiated after a run of successes with the Gemini series. There had been only one disaster, the fire on the first Apollo, on the ground, and the rather simple causes of this had been found and remedied. They knew space travel was safe. Their bias was reinforced by the fear that, if something serious had gone wrong, this moon mission would have to be abandoned.

Considering the technological, design and planning expertise that had been devoted to the Apollo series, it was surprising how difficult it was to know what was going on in the spacecraft. The practice was for Houston to do the decision-making, so there were far more instruments on earth than on the spacecraft itself. These instruments received their information by telemetry, which could be erratic, and in any case could pass on only information that the designers wanted to include. As it happened, the system *had* communicated the dramatic increase in pressure on the no. 2 oxygen tank, but Liebergot at Houston was too preoccupied to notice and be alarmed. And, by now his instruments at Houston showed that there was also an electrical fault to find. There were a lot of dials to check, but eventually the astronauts located the fault: one of the two main electrical outputs in the service module was losing power.

Then the electrical-warning light went out. Everyone felt happier and started to discuss the possibility that it wasn't the power supply that had broken down, but the instrument recording the power supply. There was indeed some justification for doubting the Apollo instruments which had, over the years, fairly frequently recorded non-existent faults. Another technician at Houston reported that there was something technically odd about the radio signals from Apollo 13, but that did not cause undue concern. Nor did the readings on the gauge of the no. 2 oxygen tank, which had returned to showing over-full.

Then Lovell reported that there was no power at all from one of the two supplies in the service module and that the power from the second was failing. This was extremely serious. There was essentially no other power-supply for the spacecraft except for three batteries that were destined for use during the last few minutes of the return to earth, after the service module had been discarded. Without power the spacecraft would soon be 'dead'. So, too, would the astronauts.

Liebergot, at Houston, had to locate the reason for the power failures quickly, and either suggest a way of curing them or find an alternative source of power. The power-supply came from three fuel cells. Two of them, nos. 1 and 2, fed one mains outlet – the one that was still working; no. 3 fed the second, which was by now dead. The fuel cells were supplied with hydrogen from two tanks that were linked together, and oxygen from another two tanks, also linked. Liebergot quickly discovered that fuel cell no. 3 had failed, and that the supply of power from the other mains was failing because fuel cell no. 1 had broken down. Then he found that even the supply from fuel cell no. 2 was failing. However, as no. 2 cell was working, even if badly, there seemed to be no reason to look at the supplies of oxygen and hydrogen. In any case, the first priority was to ensure that it wasn't simply the connections from the cells that had failed. There had after all been a bang, and this could simply have shaken the switches loose. The astronauts couldn't actually go down into the service module to look at the connections, so they operated their remote-control switches a few times in a futile attempt to get power

---

View of damaged and jettisoned Apollo 13 service module, photographed from the command module: an entire panel of the module was blown away by the explosion of the second oxygen tank

flowing. When this failed, they connected one storage battery to the one mains supply that was still working. They couldn't really spare the battery – it was needed for re-entry – but they had no choice.

The situation was further complicated by the fact that, ever since the bang, the spacecraft had been wobbling uncontrollably. Apart from the discomfort this involved, it might also lead to disaster. While the spacecraft was in flight, one side of it faced the sun and the other was in shadow. The sun-facing side could reach a temperature of 250°C, the dark side −250°C. Either extreme was harmful for the instruments on board, including those needed for navigation. The technique for achieving gentle, uniform heating was to rotate the spacecraft in what NASA calls the 'barbecue mode'. There was no way of putting the wobbling spacecraft into the barbecue mode, and there was even a risk that the wobbling would become so violent that it would actually jam parts of the navigation system.

The principal problem was still the electricity failure, so Liebergot and the crew instituted a check of the individual components of the system. Lovell quickly reported that the no. 2 oxygen tank, the one that had been troublesome all along, was now reading empty. This, if true, would explain the electrical failure because the tank fed the fuel cell. There was no way of examining the tank itself, but Lovell did float off to a window to look at the outside of the service module. He couldn't see much, but he noticed a stream of vapour rushing out into space: the oxygen was leaking away. This also explained the wobble: the leak was acting like an erratic rocket, unbalancing the craft in the same way as a child's balloon is thrown about if you blow it up and let it go before tying the neck.

When NASA technicians reconstructed the disastrous sequence of events later, they discovered that the trouble had started during a planned preliminary to the launch when the oxygen tanks had been filled in a simulation of the countdown routine. From then on, faulty design and various human errors had turned the oxygen tank into a bomb. At the end of the preliminary test the oxygen had been trapped in tank no. 2. To get it out the engineers had switched on the heating coils – the one used later for a cryogenic stir – and left them on. There was a safety switch to stop the tank from overheating, but the safety switch itself wasn't protected. It was designed to work in the cold temperatures of space: the engineers had reckoned that, in testing, it would be kept cool by the liquid oxygen in the tank. But once this had boiled off there was nothing to keep the switch cold. It fused solid, still in the on position. Everyone, fatally, relied on the equipment. No one noticed that the heating elements went on and on heating what soon became an empty container. The temperature rose and the insulation melted off the wires of the heating coil. Now, if it was switched off and then on, a spark would fly between the wires.

This is exactly what happened when Liebergot asked for a cryogenic stir on 13 April: the result of this innocent, orthodox command was a spark that heated the oxygen, building up pressure inside the tank. The tank was double-layered, with inflammable insulation packed between the walls. As the temperature rose, the tank split, the insulation caught fire, the temperature went still higher and the whole container burst. The shock – the bang heard on board – jammed shut the taps leading to fuel cells 1 and 3 and it destroyed a large section of the service module. Because the oxygen tanks were interconnected, the oxygen from the first tank leaked away through the hole in the second. The discrepancies in the radio signals noted at this time were caused by the aerial flapping about: it was attached to the piece of the module that was blown loose in the blast.

This analysis of the accident, and the re-design of the spacecraft to avoid recurrences, were still in the future. The first priority was to save the astronauts. The most immediate problem was to work out how the spacecraft would re-enter the earth's atmosphere. The crew had to be in the command module, with its heatshield, and it had to strike the atmosphere at the correct angle. Too steep, and the astronauts would be fried, despite the heatshield; too shallow, and the module would bounce off, perhaps into an enormous elliptical orbit, perhaps into space. That problem, though, could be solved only when the craft was approaching re-entry, provided that enough fuel had been reserved for manoeuvring. The controllers at Houston had to give some immediate thought to the direction of the module. The astronauts would have to be picked up once they had hit the sea, and here the successes of previous Apollo trips caused problems. At one time there had been such a large element of chance in space trips that NASA had scattered the oceans with ships – twelve or more – so that one of them would be within a few hundred miles of any possible splashdown. Because space flights had now become relatively predictable there were only four planned landing areas and, of these, only one had a ship at the ready. After all, NASA had argued, an aircraft and frogmen could arrive anywhere within a few hours. No one had considered spacecraft that would require an immediate rescue operation. Obviously, the sooner the astronauts returned to earth, the less likely they were to exhaust their supplies of water, oxygen and energy; but the longer the delay, the more chance there was of getting a ship in place to meet them. This reasoning persuaded the NASA officials to send Apollo 13 around the moon. There were various ways, expensive in energy, of turning it on its tracks and sending it straight back which would save time and get the astronauts near earth, at least, more quickly.

But they might arrive too quickly for rescue and they might have to jettison the lunar module to save energy.

NASA's plan was to use the lunar module as the 'survival vehicle', so it had to be prepared – 'powered up' in NASA language. Perhaps surprisingly, no one had planned using this module as a lifeboat: all available plans were for preparing it for a moon landing. Normally it would take a couple of hours to work through the various stages of preparation, but NASA reckoned that the astronauts had only a quarter of an hour before the power from the service module would stop altogether.

There were two immediate problems. The lunar module's navigation system had to be reset for accuracy from that in the command module. And the lunar module's oxygen-breathing system had to be activated quickly because the oxygen in the service module would run out at any moment. The astronauts crawled through the connecting tunnel into the lunar module, turned on the oxygen and calibrated the navigation system. The service module's power-supply stopped. The three men were now in the space age's first lifeboat, shipwrecked nearly a quarter of a million miles from home.

The lunar module was directed into an orbit course that would take it round the moon and back towards earth. 'Towards', at this time, was an accurate description. Unless the course were changed again, the craft would miss the earth by a couple of thousand miles. There would have to be at least two course corrections. The first rocket burn would take place a couple of hours after the craft had rounded the moon; the other as late as possible, when it became clear how much extra correction was needed. Although after the first burn the craft would be free-wheeling and should accurately hit the point aimed at on earth, its path might be disturbed by, for example, another leak of gas. It would even be disturbed if the astronauts used the rubbish vents: everything had to be packed into plastic bags and stored.

The first correction had to use the lunar module's rocket engines; it took place accurately and on schedule, although there was a slight ripple of alarm when one astronaut noticed that the switch that would discard the rocket motors – they had been designed to be left on the moon – was still on. Now the assemblage was directed at the Pacific Ocean, so NASA started arranging for a ship to meet them there.

Survival in the spacecraft was going to be difficult. Like any shipwrecked crew the astronauts were worried about water supplies. They decided to transfer some from a storage tank in the command module. Nothing in the spacecraft's design had allowed for this, so they invented a remarkably low-technology technique, bailing the water out with plastic drinking bottles. The problem with the air supply was more difficult

to solve. The lunar module contained enough oxygen for the trip, but there was concern about the potentially lethal build-up of carbon dioxide. The air on the spacecraft was circulated through tubes containing lithium hydroxide, a chemical that absorbs carbon dioxide, but as no one had considered that the lunar module might contain three people, all exhaling for four days, there would not be enough lithium hydroxide to keep the air pure. There were plenty of other unused tubes in the command module but, again because no one had foreseen this situation, these could not be fitted to the lunar module's purification system. Eventually this problem was also solved by a crude construction of plastic sheets and sticky tape to make adaptors.

Once the craft was round the moon and on its way, with a quarter of a million miles to go, there were three further problems. First it had to be put back into the barbecue mode and, by this time, a computer would have to do it because the astronauts needed rest. Secondly, an arrival method had to be worked out. The command module's batteries were designed for a usage of about three-quarters of an hour: in this emergency they would have to last for several hours, and one of the reserve batteries had already been used. For safety the command module's power would have to be nursed until the last possible moment, so the lunar module would have to provide a lot of the power for manoeuvring. Finally, the route back had to be chosen so that when the lunar module was finally discarded it would drop into very deep ocean water, because it contained a large can of radioactive material, originally intended to power experiments on the moon.

First the craft had to be directed safely to the earth's surface: although this had seemed to be going according to plan, gradually doubts had started up again. The spacecraft should have stayed on course once it had been put there, but it had developed a wobble in its barbecue spin and it was slowly going off course. If not corrected, it would hit the earth's atmosphere at too shallow an angle: it would bounce off. It was not simple to decide when to make the correction. Meanwhile, another group of engineers had noted that pressure was increasing in a helium tank – part of the rocket system in the lunar module. Eventually a safety valve would blow, and although this was designed in such a way that it should not have any effect on Apollo's flight, no one was prepared to rely on this. The sudden rush of helium might well act like a rocket. It would therefore be better to correct the course after the pressure had been released, but time was short.

Then there was a bang, and a shower of white particles from somewhere near the helium tank. But it turned out that the helium was still there. Another safety switch, for one of the batteries this time, had failed. As a result, the battery had blown liquids out through a safety valve, and

FACING PAGE
above, *Jim Irwin of the Apollo 15 crew tends the moon buggy under the shadowy mass of Mount Hadley*

below, *The Apollo 15 landing vehicle on the moon's surface; Jim Irwin salutes the US flag* the liquids, like any other liquid in space, had turned into ice and snow. The helium tank still hadn't let off its gas, and eventually the course of the spacecraft had to be corrected without waiting for it to do so: the craft was now approaching the earth very quickly. The effect of gravity is such that a spacecraft from the moon is eventually travelling at 24,000 mph as it reaches the earth's atmosphere. The stricken Apollo 13 was already travelling at thousands of miles per hour.

The astronauts were beginning to suffer. Because they were aware of the need to reserve water in order to keep the machinery working, they had been secretly rationing themselves, ignoring the directions from Houston. As the effect of weightlessness is to diminish the feeling of thirst but not the need for water, they were becoming dehydrated. Haise had developed a kidney infection which was worsening, and they were also becoming very cold. They were using the command module as a bedroom, but the heating equipment had been shut down and the temperature there was near freezing. The sleeping-bags were rather nominal – they were intended to prevent the sleepers from drifting round the spacecraft, not to keep them warm. So the astronauts slept badly because of the cold and felt colder because they had slept badly. In addition, although they were experienced, trained men, they were un-doubtedly worried about the risks of landing. They were worried, too, about the technical problems of getting power from the lunar module to the command module – everything had been designed to send power the other way. Then the team at Houston discovered that the spacecraft had once again drifted into a shallow path that would bounce them off the atmosphere. Another course correction would be needed.

There were now only eighteen hours to go. Houston had been working out a procedure for getting the spacecraft down, jettisoning first the service module and then the lunar module. Because no one had pre-viously envisaged this routine they had to devise an enormously detailed plan, listing in order every switch that had to be moved, every valve that had to be opened or closed, every dial that had to be read. The plan was to be run through a computer that could simulate the procedures and check that they were adequate, then it was to be read to the astronauts. It was not surprising that the astronauts became tense and edgy.

On the most important night of their lives, Lovell, Swigert and Haise managed only a couple of hours of sleep before the cold and the worry aroused them: there was a real risk that they would make mistakes in landing simply because of tiredness. There were supplies of Dexedrine, an amphetamine stimulant, in the spacecraft, but these gave them no sleep, simply made it easier for a while to work without it: when the effect wore off, all the postponed fatigue would suddenly return. NASA advised the astronauts to take the pills, but they postponed doing so until they were sure there would be nothing significant to do after the effects had worn off.

When the spacecraft reached a point 58,000 miles away from earth, it was travelling at 5900 miles per hour and accelerating rapidly. It was time to start preparing for arrival. At around 5 a.m. Swigert started using electricity from the lunar module to heat the instruments and the control thrusters on the command module. These had been allowed to cool down to save energy while the astronauts were living in the lunar module. Now they had to be readied: only the command module had a heatshield for re-entry. The spacecraft had again drifted from its course and inexplicably seemed to be moving even further away. Another correcting rocket burn was needed.

The astronauts, out of fatigue and dehydration, started making mistakes. An instruction had been wrongly copied: this would have led to using the wrong rockets, but the error was noted and corrected on earth. The spacecraft was put into the wrong attitude for the burn; again the error was noticed from earth and pointed out to the astronauts. Eventually the burn was made and the spacecraft was put back into the centre of its 're-entry corridor'.

The operations were now critical. There was, at that moment, no telemetry to earth because of lack of power, and the astronauts had to jettison the service module by breaking the connections with small charges of explosive that had been in position since launch for that purpose. The charges were detonated and the lunar module's thrusters fired in reverse so that it slowly drew back from the service module, taking the command module with it. Similar charges and similar switches would later separate the lunar module, which was, at this stage, still necessary for survival.

As the service module moved away, the extent of damage became visible: it was the first and last time anyone was to see it. The astronauts were appalled. They saw that the service module had split apart: the oxygen tanks were at the near end and had clearly burst violently. They suddenly realised with horror that the command module's heatshield might have been damaged as a result.

There was, in fact, no point in worrying about the heatshield: whatever its state, there was nothing that could be done. The immediate problem was the power-supply. The only power in use, and the only communications with Houston, came from the lunar module. It was time to think about connecting the command module's re-entry batteries. One battery was connected experimentally to check the problem. To the alarm of Houston, a current flowed. A switch had been left on: if uncorrected, this would drain the battery. The telemetry from the command module wasn't working, so there was no way of locating the faulty switch. The battery was, for the time being, disconnected as an emergency measure: there was no electricity to spare. Then, at two and a half hours from splashdown, all the batteries were connected and the power from the lunar module cut off.

Two and a half hours from splashdown is only two hours from re-entry. By re-entry the command module had to be working fully, its computer loaded with the correct program, its navigational instruments aligned. The lunar module by then had to have been jettisoned. There was a lot to do. Quite early the telemetry was switched on and the problem switch located. The astronauts were told to turn it off, but they weren't told why. It would be unsettling to be told about errors.

Another urgent priority was the setting of the navigational equipment. Despite all the technological sophistication of a spacecraft, it was still best to use the stars to establish its exact position – to 'get a fix'. But the stars couldn't be seen because the spacecraft was accompanied by a cloud of white chunks of ice, formed from the water from the spacecraft's cooling system. Then NASA decided to use a star on the shadow side of the spacecraft where the glare from the ice particles should be less. This method was eventually successful, and the astronauts were able to see whether errors had crept into their navigational equipment in the few hours since it had been aligned with the set on the lunar module. There were none.

They still had to detach the lunar module, using that module's rockets as a final delicate corrector to get into the proper alignment. Lovell, as Commander, took on the task – another that had never been performed before this trip. Eventually the alignment seemed correct and Lovell scrabbled through the connecting tunnel for the last time, leaving the lunar module crammed with bags of rubbish and sewage.

He had made a serious mistake: the alignment was inaccurate by 90°, meaning that the lunar module would go off in the wrong direction, and the command module would be thrown off course in the opposite direction. As it was too late to return to the lunar module and re-orientate it, Mission Control decided to accept the inevitable: the path of the command module would have to be corrected later.

There were still further problems. The plan was to keep the pressure in the connecting tunnel high, so that the two craft would be blown apart once they could separate. But the only way to test the airtightness of the hatch on the command module was to let some of the air out of the tunnel, to see what leaked past the hatch. After a moment of alarm, when there seemed to be a leak, the NASA engineers decided that the hatch was sealed. By now, the lunar module was becoming a nuisance – it was throwing the combined craft off balance. The remarkable lifeboat was finally jettisoned.

For the first time since the original accident Apollo 13 returned to its normal routine. The three astronauts were in the command module, as they should have been, speeding towards the edge of earth's atmosphere and then splashdown. They still had 11,000 miles to go on a route that would take them around the earth, to drop into the atmosphere and then on down towards the aircraft carrier *Iwo Jima*, which was waiting for them in the Pacific Ocean. It was now falling like a partly controllable stone. There should be no more problems.

Apollo 13 astronauts – John L. Swigert, Fred W. Haise and James S. Lovell – are greeted by President Richard Nixon after their successful splashdown and recovery

But there were. Again, the craft was found to be inexplicably changing its course, getting nearer to skimming away. As no one knew what had caused the drift, no one could guess if it would continue. It had a mere tenth of a degree left if it was to stay in a safe corridor. The spacecraft might need a change of position but no one wanted to do this unnecessarily. The astronauts weren't even told of the problem, but they knew there was one because NASA wouldn't send the timetable for the last stages of the descent, which would be affected by any changes. In the end, Mission Control decided that the course would be safe: they realised that it was the water boiling off the lunar module's cooling system that had caused the drift. Once the module had gone, so had the drift.

Suddenly, and at last, everything started to go right. The command module flamed as predicted through the atmosphere and the astronauts, after a lengthy period cut off from radio contact, came down in the ocean at 12.07, Central Standard Time. Soon they were on the *Iwo Jima* and the party started, but without its main actors: the Mission Control people were tired and the astronauts were in the sick bay. They were extremely weak; they were suffering badly from the effects of weightlessness; and they were severely dehydrated. They had only just made it.

7

UNMANNED PROBES TO THE PLANETS

Long before men landed on the moon there were dreams of going further; and after the landings the ambitions of the space enthusiasts soared. Would it be possible to reach Mars, a planet where life might exist, and what were the prospects for more distant planets? And then there were plenty of stars, many of them with planets, and some of these planets would be similar to the earth. Could we send people out to look at these?

We certainly know how to send a spacecraft to Mars and land it automatically on the surface, but these spacecraft carry only instruments, not people: there are immense problems about landing people because of the massive life-support systems that are required for such very long journeys.

All distances in space are so enormous that it is difficult to have any conception of the relation between them. The moon is a quarter of a million miles away: Mars, at its closest, about 50 million, i.e. 200 times as far. It took approximately two and a half days to reach the moon – and the whole journey from blast-off to landing took four days. It would take 500 days – sixteen and a half months – to reach Mars. Provisioning for a trip that long, when oxygen and fuels must be carried as well as food, involves an enormous spacecraft.

Although Venus is somewhat nearer than Mars, the other planets are more distant. There are nine planets including earth, but only five of them can be seen without a telescope. All planets travel around the sun in ellipses that are very nearly circles. They travel at speeds that balance the gravitational pull of the sun inwards – there are other forces acting on planets, but these are much smaller. To balance this gravitational pull, the innermost planets move at great speed, while the outer ones move in more leisurely orbits. Mercury, the innermost, takes eighty-seven days per orbit: Pluto, the outermost, just over 248 years. The average distance of Mercury from the sun is 36 million miles: that of Pluto, more than a hundred times that – 3670 million.

Although the five planets that can be seen with a naked eye have been seen from the beginning of recorded time, and presumably earlier, none of them is luminous. All the planets shine, like the moon, by reflected sunlight. They were all formed, together with the sun itself, about 4500 million years ago from a vast cloud – a primitive solar nebula. This cloud was disturbed, perhaps by a passing star, so that it broke up into dense patches. The massive hot central part became the sun: the planets condensed outside. Those nearer the centre – Mars, earth, Venus and Mercury – are relatively small and dense (see endpapers), while the outer ones, being constituted mainly of volatile substances that would be boiled off a planet nearer the sun, are large and have a much lower density. Saturn would even float on water if it could be placed in a tank large enough.

The sun is a typical star, like the others we see at night, but we see the sun as a ball while we see the others as points because they are so much further away. We can get a feeling for the relative distances of planets and stars by remembering that no telescope ever made has shown a star as more than a point: we cannot imagine a telescope so powerful that a star would appear as a disc. But even a modest telescope shows the planets as discs and can reveal the moons travelling around other planets, just as our moon travels around us.

The sun is about 94 million miles from earth. The whole solar system is a modest 9000 million miles across. This distance can be called modest because our nearest star (actually a group of three stars referred to as Alpha Centauri) is more than 25 million million miles from us. In other words, it is 30 million times as far away as the sun. Other stars are much more distant still. The Milky Way is a wide belt of closely packed stars, and on either side of this the stars are more scattered. The stars in the Milky Way are not really packed more closely together than in the rest of the sky – they simply look like that because we are looking at a deeper layer of stars, spread out evenly. Most of the stars we can see are arranged within the pattern you would get by placing two saucers together, rim to rim. The sun and the earth are very roughly half way between the centre and the edge of one rim. When we look at the Milky Way we are looking across the saucers. We see lots of stars, apparently clumped together. When we look at other parts of the sky we are looking through a thinner layer of stars. This whole double saucer of stars is called a galaxy. There are other galaxies as well. The centre of ours is roughly 195,000 million million miles away.

The double saucer is about 588,000 million million miles across and, at its centre, more than 17 million million miles thick. In this enormous volume there are between 100,000 million and 200,000 million stars, in fact only a thin scattering. Of the other galaxies, the nearest, called M31, is 12 million million million miles away. It is called M31 because it was No. 31 in a catalogue of nebulae – many of which are galaxies – published by the French astronomer, Charles Messier, in the eighteenth century. They all contain approximately the same number of stars.

Some of these stars have planets. While it isn't conceivable that there could be life on a star, there could be life on some planets even if they are different from planet earth. There must be some inhabited planets – planets where the conditions, by chance, were at some time conducive to the development of life, just as they were right on earth $4\frac{1}{4}$ thousand million years ago. However, communication with other inhabitants of space is impossible because of the distances involved.

For convenience, astronomers use a different unit from the mile to describe distances: this unit is the light year – the distance that light would travel in a year, i.e. 5,880,000 million miles. Working in light years, it is possible to use reasonable numbers in describing distances in the universe. Alpha Centauri, for example, is 4·3 light years away, a less cumbersome number than 25·284 million million miles. Our galaxy is 100,000 light years across and 20,000 light years thick at the centre, 3000 light years thick at the edges.

THE SOLAR SYSTEM

The solar system comprises the sun and the nine known planets, plus the collection of orbiting rocks known as asteroids and also the comets. In drawings, the tracks of the planets and asteroids are usually shown as ellipses round the sun (the comets loop in very elongated ellipses sharply inclined to the plane of the planets), but these drawings mislead. Even when the orbits of the planets are drawn to scale, the planets are not. And they don't occupy the orbits: they merely move around them. One can easily fail to gain an accurate impression of the vast areas of empty space.

The comets stay in orbit around the sun, invisible until disturbed by some passing body that sends them towards the sun. A few go into elliptical orbit, so that they reappear regularly – Halley's comet, visible approximately every seventy-six years, is the most famous example. Most comets appear only once and then disappear forever into space. Their paths may be elliptical for some of the time, but the matter is academic as they are not repeated. The 'abode of the comets' is part of the solar system, but as we cannot observe them while they are there, and cannot predict when they will be disturbed, there is little point in further discussion.

The solar system has a total diameter, logically enough, of the orbit of the most distant planet – Pluto. Pluto's average distance from the sun is about 3700 million miles. Like all the planets, Pluto moves round an ellipse, not a circle: unlike the other planets, its ellipse is quite strikingly non-circular – when Pluto is at its nearest to the sun, it is actually closer than Neptune, the next planet in. It is a tiny planet – only 2500 miles across – about a fifth of the size of the earth. It is probably simply a tiny ball of solid methane, mixed with a few other substances – simple compounds of carbon, hydrogen and nitrogen – that would all disappear if it were nearer the sun.

The average distance between Pluto's orbit and that of Neptune is about 875 million miles. Surrounding Neptune there is empty space – looking outwards from the sun. The only object between Neptune and the nearest star is Pluto, a modest 1700 miles across. Neptune's average distance from the sun is 2800 million miles. The planet has a density of 1·7, which means that it consists of rocks as well as ice. It is about 31,000 miles in diameter – nearly four times as big as the earth. Like the earth, Neptune produces heat internally. The earth's heat comes from radio-activity, and so presumably does Neptune's, although the planet is too far off for us to know this for certain.

There is another enormous gap – an average of more than 1600 million miles – between Neptune's orbit and that of Uranus. This planet is larger than Neptune, and it has a density of 1·2, only slightly higher than that of water. Like the other outer planets it has a core of rocks, surrounded by ice and volatile substances – methane in particular. Uranus is unlike any other planet in two ways: it has no internal heating, and it spins, once every sixty-five years, on an axis pointing to the sun.

The planets travel around the sun in ellipses, spinning so that every part sees sunlight daily. Some planets, including earth, spin at a slight angle to the ellipse, so that the north and south poles have six-month days and six-month nights; however, these polar regions form only a small part of the earth. Uranus spins in such a way that one pole is always dark, the other always lit. Because the axis does not point exactly towards the sun, there are some places that experience a day followed by a night. In this case, the total time from daybreak to daybreak is sixty-five years.

The three outer planets – Uranus, Neptune and Pluto – are relatively new discoveries, first observed in 1781, 1846, and 1930 respectively. The other five planets have been charted since the beginning of recorded time, because either their proximity or their size made them easily recognisable.

At a distance of 94 million miles from the sun, the earth has a density of 5·5, the highest among the planets. This is because it has a core of liquid metal – the swirlings of this liquid metal produce the earth's magnetic field. Its crust contains radioactive substances that generate heat – sufficient heat to melt the rocks under the surface – and this molten rock occasionally spouts forth from volcanoes. In addition, the crust of the earth is not fixed in position. Great plates of crust drift on the surface, producing earthquake zones and chains of volcanoes where they collide. This doesn't seem to happen with any other planet.

The innermost planet and the smallest is Mercury. It is only 3000 miles in diameter – not very much larger than the moon – and it travels on average, only 36 million miles from the sun. This tiny planet appears only as a morning or an evening star, and even at those times it is difficult to see – it is near the horizon, often obscured by haze.

Mercury, like Venus and the earth, has a high density – 5·4 – which suggests that it has a metal core beneath a crust of silicate rocks. It has no atmosphere – there are no gases at all above the surface of the planet – because it is too near the sun to retain one, so its surface temperature varies widely, from over 430°C in the hottest parts during its day to below −173°C at night. And because it has no atmosphere its surface is pitted with craters produced by the impact of meteorites arriving from space.

The planets of the solar system – Mercury, Venus, Mars, Jupiter, Saturn, Uranus, Neptune and Pluto – are our nearest neighbours and the only celestial bodies, apart from the moon, that we can realistically hope to visit by spacecraft in the near future. By studying them we can gain greater understanding of how the earth came to be formed – of how life arose on earth perhaps. It isn't surprising that space scientists have planned ingenious unmanned space trips to examine the planets more fully than we can do from earth, and the discoveries are fascinating. They have so far explored the planets from Mercury out to Saturn.

MARINER GOES TO MERCURY

Manned space flight captures the imagination, and for some practical purposes there is no substitute for a person in space. An astronaut can react quickly and flexibly to circumstances, as demonstrated by the salvaging of Apollo 13. And it is obviously true that a human being can recover and repair an ailing satellite where an unmanned spacecraft would be useless.

But these advantages are in near space. The Shuttle orbits, say, 120 miles up (although it can go as far as 500 or so miles up); a communications satellite is positioned 22,300 miles above the earth. The nearest planet, Venus, is 40 million miles away at its nearest. This is the equivalent of a hundred and sixty times the distance of our remotest

Mariner 10 spacecraft

manned space exploit – the trips to the moon. If we want more information about the planets, we can't wait until we are able, let alone can afford, to send people there: we must use unmanned probes. And because these unmanned spacecraft have to travel such enormous distances – much further than the mere distance between earth and the planet, because a spacecraft doesn't fly as a crow does – unmanned space explorations are the triumph of space research. They enable us to reach distant planets, in some cases to land the craft on their appallingly inhospitable surfaces, and to undertake research into their origins and histories. The planetary probes are a triumph of elegant technology.

The Mariner explorations of Mercury are probably the most elegant of these experiments. Mariner 10 succeeded in producing detailed pictures of a planet that is extremely difficult to observe at all by telescope.

Mercury, at its nearest, is about 57·5 million miles from the earth. On 1 November 1973 Mariner 10 was launched towards Venus at a speed that was carefully computed so that the gravitational effect of Venus would send the probe around the sun, meeting Mercury every lap at the same place in the planet's orbit. The probe might have stayed in orbit around the sun for ever, sending back regular reports from Mercury: in fact it made three encounters, on 29 March 1972, 21 September 1974 and 16 March 1975, and then stopped transmitting.

104

Before these observations, very little was known about Mercury. The planet travels around an ellipse at an average distance of 36 million miles from the sun and it is unique among planets because the ellipse slowly turns through space (see endpapers). This very peculiar behaviour could not be explained before Einstein put forward his relativity theory.

Another peculiarity of Mercury's orbit was discovered in 1965. There is no way of telling from observations by telescope how quickly Mercury spins, because there are no clearly recognisable landmarks on the planet. But radar showed, in 1965, that Mercury rotates once every 58·6 days: this equals two-thirds of the 88 days it takes Mercury to travel around the sun. In other words, Mercury makes three turns on its own axis for every two laps of the sun. This cannot be an accident: it must be an effect of gravity. Like the rotation of the moon, of which one side faces the earth continuously, there must be a physical reason and it must be connected with the mass of the sun. The rotation, incidentally, produces some very odd heating effects on Mercury, because particular places on the planet's surface stay stationary under the sun for quite long periods. If it were possible to stand at the right place on Mercury's surface, one would see the sun apparently stop in the sky, then go backwards, and finally re-start.

These radar observations could not provide much information about the physical nature of the planet, but Mariner 10 could. The probe carried a device to measure temperatures on the surface of Mercury, a spectrometer that would identify gases in the planet's atmosphere if it had one, and TV cameras which photographed the planet in striking detail.

Mercury is dead – its proximity to the sun makes it uninhabitable. And it is covered with craters, so that it looks much like the moon. Rather surprisingly there are a number of smooth plains, some of which occur in a cratered area. The craters were made by meteorites hurtling in from space, and because these varied in size, so do the craters. Craters have frequently been covered by others which formed later, but the earlier craters can often still be identified – they are less clearly defined than the later ones because the edges have been eroded. The probes found no trace of any kind of atmosphere: the erosion on the planet has been produced by a rain of microscopic particles – micrometeorites – from space.

The most striking single feature on the surface of Mercury is the gigantic Caloris Basin. This is a circular plain, 8125 miles across, surrounded by a ring of mountains two kilometres high. The extraordinary fact is that these mountains were not formed, as the mountains on earth were, by pressure that crumpled the crust. The plateau of Caloris was

Photograph of surface of Mercury showing impact crater, taken by Mariner 10 from a distance of 21,100 miles as the spacecraft approached Mercury for the first time, 29 March 1974

produced by the impact of a gigantic body from space, and the mountains were forced up by the impact. This collision was so powerful that it affected the opposite side of Mercury: there the surface consists of a broken rubble of rocks that were produced by shock waves travelling right the way through the planet.

The craters on Mercury, and the broken area caused by the giant collision that formed the Caloris Basin are very striking, but they are easily understood. Mercury has no atmosphere so every meteorite reached its surface complete. The pock-marked surface is the result. What is difficult to understand are the parts of the surface that are unmarked, the smooth plains – especially the plains between the large craters. Mercury should have been covered more or less uniformly with craters: one group of scientists believe that it was, and that some of these craters have been smoothed over by flows of volcanic lava. There is at present no way of knowing, though there is certainly no evidence of active volcanoes now. If there was a volcanic period it must have been soon after the planet was formed 4·6 billion years ago.

106

As it is the planet nearest the sun, Mercury is composed of materials that are not evaporated by heat. It has an enormous iron core – the largest, proportionately, of any planet – which was formed at the same time as the very thin layer of minerals. The bombardment that produced the large craters dates from soon after the planet's formation, and the unidentified process that smoothed out the plains between the craters occurred soon afterwards. Then, about four thousand million years ago the Caloris Basin was formed and this was again followed by the process that produced the smooth plains. Since that time, apart from the occasional small meteorite, nothing has happened to change the planet's structure.

VENUS

The earth goes around the sun approximately 50 million miles inside Mars. Venus orbits 26 million miles inside the earth. Had conditions been only slightly different, Venus might well have become another earth. The two planets are certainly very similar in their physical characteristics. The diameter of Venus is slightly less, 7565 miles, compared with 7972 miles, and its mass also less: $4·9 \times 10^{24}$ kg to the earth's $6·0 \times 10^{24}$. So the densities are close – Venus, 5·2; earth, 5·5. At its closest, Venus is a modest 25 million miles from us. It is our nearest planet.

In every other way, however, it is dramatically different. Venus spins in the opposite direction to all the others, once every 243 days, but its atmosphere rotates once every four days: the atmosphere is moving much faster than the planet beneath. In other words, Venus has steady winds at enormous speeds – around 225 mph – which blow at the top of the atmosphere: near the ground there are only gentle breezes. The temperature at ground level is around 477°C, the pressure ninety times that at the earth's surface. The atmosphere is ninety-seven per cent carbon dioxide, but the surface of the planet seems to resemble ours. The rocks there are probably similar and they are distorted, as ours are, by internal forces. But there is of course no vegetation – no life of any kind.

The earth's atmosphere was originally mainly carbon dioxide which was converted by the first green plants to the oxygen that animals need to breathe. Life could have started on Venus if green plants had happened to grow there, but at the moment it is a hot, dry, inhospitable planet. It is the easiest of the planets to recognise as it is seen as a brilliant morning or evening star.

Before the advent of the space age Venus was one of the most mysterious planets: its surface was invisible, blanketed by cloud, and we knew nothing of its geography, let alone its geology or history. Because its size and mass were so similar to those of the earth, it was talked about

107

as an earth that had never happened: a desert on which the sequence of chances that had led to life on earth hadn't occurred.

The first spacecraft visits to Venus revealed how different it actually was from earth. It is such a hostile planet that these visits merely highlighted the problems a successful investigation would have to overcome. The visits, modest at first, but eventually successful and valuable, were made mainly by Russian spacecraft.

Venera 4, successor to a series of three failed attempts, arrived at Venus on 18 October 1967 and descended by parachute. It sent signals back to earth for only 94 minutes. As it descended its instruments recorded temperatures reaching 271°C and an awe-inspiring pressure of about twenty times the atmospheric pressure on earth. The US probe, Mariner 5, went past Venus at much the same time and established that Venera 4 had not discovered the temperature or pressure on the surface of Venus: it was still 15 miles above the surface of the planet – the height of three Everests – when the extreme conditions destroyed its transmitter.

As space technology advanced in the late 1960s it wasn't very difficult to reach Venus with a probe – the Soviet Venera 5 and Venera 6 both reached the surface of the planet. The problem was to design instruments and radio transmitters that could work at the temperatures and pressures there. Venera 7, carefully shielded, managed to operate on the surface, but only for a quarter of an hour. The messages it sent back revealed that it was operating at a temperature of 477°C – red-heat – and a pressure of 1350 lb per square inch (ninety times our atmospheric pressure). Venera 8, in 1972, lasted for 50 minutes, and it reported the slightly surprising fact that, despite the clouds that cover Venus, some sunlight reaches its surface – enough, as later visits were to confirm, to make TV pictures possible.

So the next Russian probes, sent to Venus in 1975, carried TV cameras and refrigerating equipment. Venera 9 and Venera 10 were complex two-part craft: one part was designed to orbit the planet, sending back the information that could be discovered from above; the other was a soft-lander.

The soft-landing system was ingenious. It needed a parachute, but the Russians realised that a full descent by parachute might take so long that the scientific equipment would be baked before it reached the surface. The probe therefore slipped its parachute while still well above the surface. It was shaped so that it fell only relatively slowly through the dense atmosphere, but much faster than a parachute descent.

Venera 4, the Soviet automatic satellite which made the first controlled descent on to the surface of Venus on 18 October 1967, and relayed data on the planet's atmosphere

Each probe successfully sent back pictures, and these were impressive firsts: before this no pictures had ever been received from another planet. (The Mariner visits to Mercury were later, in 1974.) From Venera 9's pictures Venus seemed bleak, with rocks scattered all over the panoramic view, but Venera 10 landed 6000 miles away and its pictures showed a smooth rock covered with granules. The pictures from the probes show only small areas, and because they are panoramic pictures they are sometimes difficult to interpret. However, a great deal can be deduced from one picture, as is the case with one of Venera 9's shots. The camera is actually looking up a steep slope, and the rocks in the picture are large with rounded edges, having clearly rolled to their present position. This is a scree slope – a stack of rocks that have been set free of a mountain by erosion and have fallen down the slope. The existence of such a slope implies what geologists call activity: erosion implies an atmosphere.

Venera 10's camera looks out from the top of an outcrop which appears to have been fractured: the fractures show steps, which on earth would be an indication of a sedimentary rock. On earth these are generally produced at the bottom of oceans, but there is no reason for believing that there were ever oceans on Venus. Sedimentary rocks can also be formed on land, where winds stack up pulverised material. As the atmosphere on Venus is very dense – ninety times that of earth – and the winds at the top of the atmosphere are very powerful, these facts might possibly explain the erosion and the formation of sedimentary rocks. However, although the winds at the top of Venus's atmosphere are very powerful, with speeds of more than 225 mph, the probes found virtually no wind at all in the six miles nearest the surface. It would be wrong to assume that a few measurements can indicate the average winds on the whole planet, but on the other hand it would be absurd simply to assume that there are usually high winds and that the probes happened to land on unusually calm days.

From a basic knowledge of meteorology it is possible to predict that Venus will generally have calm, boring, though appallingly hot weather. Its axis of rotation points directly away from the sun, so there are no seasons. It rotates only once every 243 of our days and it has a very dense atmosphere, so the temperature stays much the same, day and night. It is as hot at the poles as at the equator.

Erosion on Venus doesn't therefore have the same causes as it does on earth: the only reasonable proposal is that somehow dust, sand and grit are thrown into the Venusian atmosphere. The density of the atmosphere would allow these granules to collect and run in streams down any slopes on the planet. Another possibility is that the pictures don't show sedimentary rocks at all. The rocks could come from volcanoes because lavas do sometimes settle out as separate layers. The pictures from Veneras 9 and 10 are not good enough to decide, nor is there any certain evidence of volcanoes, although some pictures appear to show one.

There are also craters - enormous craters - on Venus, and these, too, are difficult to explain. They are much too big to be caused by meteorites, although they could be volcanic. The probability is that each of them was caused by a collision with an enormous chunk of rock - so vast that it melted the surface of the planet. There are giant craters like this on the moon - the maria - but they are easier to understand because the moon has no atmosphere to shield it from collision: the atmosphere of Venus is extensive and dense.

Further expeditions have told us no more about the surface of Venus, though they have confirmed our previous knowledge of the atmosphere. In December 1978 two Russian craft, Venera 11 and 12, landed, but they failed to send back pictures.

At the same time the Americans sent up Pioneer Venus 1 and 2 which succeeded in humbler tasks. Pioneer Venus 1 was directed from earth to orbit closer and closer to Venus until it moved through the very upper atmosphere. Pioneer Venus 2 was a complex device that eventually threw out three small probes and a large one. These discovered that the clouds - the characteristics of Venus - occurred in three layers. The upper two were faint and hazy: these are invisible from earth and probably have no effect on its climate. The significant layer, the lowest one, was 35 miles above the surface and was dense. These were rainclouds, and one of the Pioneer Venus probes detected rainfall. The rain is not water but sulphuric acid which, because the planet is so hot, never reaches its surface. It evaporates and returns to the clouds.

The temperature that evaporates the rain would make life intolerable. The Venus Pioneer probes confirmed that although there were a few per cent of nitrogen in the Venusian atmosphere, and traces of sulphur dioxide, water and oxygen, the bulk was carbon dioxide. This gas has a peculiar ability to trap sunlight as heat in exactly the way that the glass of a greenhouse operates. The enormous temperatures on Venus almost certainly result from its atmosphere.

If green plants were introduced on to Venus, the carbon dioxide would be converted to oxygen and the temperature on the planet would fall. The atmosphere on earth may, similiarly, have been rich in carbon dioxide until there were green plants. There was even once a project, not really a serious one, to send a rocket-load of algae to Venus and initiate the conversion of the planet into a second earth. However, the discovery that Venus has no water, and that the temperatures there are so high that plant life isn't feasible, showed that this project wasn't practicable.

MISSIONS TO MARS

Mars travels at an average of 142·5 million miles from the sun. It is a small, solid planet, only half the size of the earth, and its density is only 3·9, compared to the earth's 5·5. For a number of reasons many of those who believe that there is life in the solar system beyond earth have always favoured Mars, but in fact none of the space probes has come up with any convincing evidence of life. However, they have found out a lot about the nature of the planet and of its two moons, Phobos and Deimos.

Every astronomer, at some time or other, has talked of life on Mars. The planet comes close to earth – 35 million miles at its nearest – and when it is that close it is very bright and recognisable: its movements have been recorded throughout history. But it is tiny – only 4242 miles across, not much more than half the diameter of the earth. Only Mercury and Pluto are smaller. So it has never been easy to see details on the planet, even with the best telescopes – especially as these telescopes have to penetrate the earth's disturbed atmosphere before they can see into space.

Because astronomers have had difficulty in seeing any fine detail on Mars they have made great mistakes in interpreting what they have seen. In particular, some of them drew maps showing canals, which implied a generous amount of water on Mars – and thence a belief that there must be canal-builders on this apparently hospitable planet. This led to one of the most influential of all science fiction novels – H.G. Wells's *The War of the Worlds*, in which Martians invade the earth. This was the story that generated a panic when it was transmitted as a radio play in New York.

The canals came into astronomical being during the nineteenth century. An Italian, Pietro Secchi, described them in 1869, though without attracting much attention. Eight years later another Italian, Giovanni Virginio Schiaparelli, mapped them very carefully, and certainly believed that they were waterways though he didn't consider the idea that they might be artificial. Then, at the end of the century, Percival Lowell, a rich American amateur astronomer, built what is still a famous observatory at Flagstaff, Arizona, to study the canals. He was sure that they had been built by intelligent Martians and he wrote books describing their feats of engineering.

The advanced telescopes and space-probe pictures of the twentieth century finally showed with certainty that there were no canals. This series of episodes demonstrates that science is not as objective as scientists sometimes claim and non-scientists sometimes fear.

When space-probes first came into existence Mars was still a potential habitat for 'life beyond earth' and was therefore an attractive planet to study. Both the Russians and the Americans failed in their first attempts

112

with probes, but on 14 July 1965 Mariner 4 'skimmed' Mars 6250 miles up and sent back twenty-two rather fuzzy pictures of a sterile, cratered surface. The possibility of intelligent inhabitants of Mars disappeared. The belief that there may be some form of life there, though, still persists.

Because the earth and Mars orbit the sun at different speeds, the distance between the two planets varies very widely: the greatest distance is about eight times the least. During 1971 the planet came particularly close to earth, thus creating a good period for probes. The Russians and the Americans each sent two. The Russian probes, intended as soft-landers, crashed. The first American probe, Mariner 8, didn't even leave the earth, but Mariner 9 was an outstanding success: it went into orbit around Mars and sent back a stream of TV pictures - more than 7000 in all. Suddenly, the whole planet could be mapped and the results were somewhat surprising.

The southern half of the planet was covered with craters, and it seems that the bombardment that had produced them had occurred soon after the planet was formed. The Northern hemisphere had vast smooth plains that seemed to have a top layer of basaltic lava, and it contained four gigantic volcanoes - one of them the largest in the solar system. It was so big that, using the best of telescopes, it could be seen as a white spot from earth. Until Mariner 9's trip it was known as Nix Olympica - the snows of Olympus - because it was believed to be a white mountain. As

Mars 3, one of the unmanned Soviet spacecraft which landed on Mars

a volcano it had to be renamed and is now called Olympus Mons.

These Martian volcanoes are what are known as 'shield' volcanoes: they are wide and gently sloped, not typically cone-shaped. Shield volcanoes let their lava out slowly, through a number of holes, over a long period. It is possible, though not certain, that this group of volcanoes is still active. Active or not, they show that Mars has not always been geologically dead, although there isn't much evidence to show how the volcanoes were produced. On earth, the grinding together of gigantic moving 'plates' on the surface of the earth produces volcanoes near the edge where the plates meet, but we don't know if this is also true of Mars. But there is what seems to be a gigantic rift valley – a deep canyon 3125 miles long, far larger than any other known in the solar system. From space-probe pictures this valley, named Valles Marineris (after the Mariner probes), looks like a large version of the Rift Valley in Kenya, which was produced by the movement of continent-sized plates.

As well as volcanoes and rift valleys, Mariner 9 made the more fascinating discovery of what seemed to be watercourses. If there was water on Mars, there must at one time have been a dense atmosphere. And if there had been a dense atmosphere, there could have been life. These watercourses deserved close inspection.

The pictures from Mariner 9 showed what looked quite remarkably like dried river-beds. They had tributaries on the high land, and they ran fairly directly down steep slopes. In the plains they meandered, and on the broader plains they sometimes formed deltas. It would be difficult to believe that they were anything other than what they seemed to be.

But if they were dried river-beds, how and when were they formed? The atmospheric pressure on Mars is so low that water cannot flow on the surface because it would evaporate. There are two possibilities: either the river-beds were not formed conventionally, or there was a time when the atmospheric pressure on Mars was much higher – high enough perhaps to support some kind of life. While there was no evidence of the mythical canals, nor of a race of Martians, this was evidence that life on Mars might once have been possible.

It is *liquid* water that is lacking on Mars. There are enormous amounts of ice, in the polar ice-caps and as permafrost underground; there is also water combined in minerals. The river-beds might have been the consequence of interaction between underground heat and ice. This would form streams. Some of the rivers appear to start in canyons of scattered rocks. These could be the remnants of a vast underground deposit of ice that was melted by underground heat to become the source of a river.

Another possibility involves an odd process that would make the river grow backwards. This sometimes occurs on earth, so it is worth considering. The process begins with a geological event – a landslide, say – that exposes the permafrost layer. The ice will melt or sublime away and the rock nearby will then crumble, exposing more permafrost, which will undermine more of the hill as it disappears. The consequence of both this and the first proposal is that water could only have flowed leaving river-like traces if the atmospheric pressure on Mars was much higher at some earlier time, with normal rainfall and normal rivers, and providing that the planet was warm enough for the water not to freeze. In other words life could, in the past, have been possible on Mars. If it did exist, there might still be traces.

The two Viking space-probes were designed to look for traces of life and they were marvellously sophisticated. Each of them consisted of an orbiter, which stayed above the planet, and a soft-lander, which was thrown from the orbiter. The landers parachuted down for the first part of the descent and were slowed by rockets for the last: it was impossible for a parachute to bring off a genuinely soft-landing in the thin atmosphere of Mars. As the landers carried cameras and soil samplers it was important that they landed gently and stayed upright. The orbiters, with the landers attached, first went round the planet for days, sending back TV pictures, so that the team back on earth could choose a suitable site. It would be difficult to land in a rocky area, or one covered with a deep layer of dust, so the lander was directed at a rather featureless plateau. No doubt there were more interesting terrains, but it would be absurd to risk failure in this immensely expensive first experiment. Once the soft-landing technique had been proven, it could be applied for more treacherous landings.

Viking 1 was launched on 20 August 1975. The aim was a soft-landing on Independence Day – 4 July – of the following year, but the surface of Mars turned out to be much rockier than expected from the Mariner 9 pictures, so the landing was delayed while the team on earth considered alternative sites. It finally occurred two and a half weeks later on 20 July, the anniversary of the first moon landing. Viking 2 landed on 3 September.

The landers set about looking for signs of life. There was no question of looking for animals or even plants in the usual sense, but if Mars had at some time been somewhat similar to earth, then simple forms of life – single-celled creatures, even bacteria – might have come into being. And if they had been formed, they could have evolved in such a way as to survive as the conditions on the planet changed. So each lander had a mechanical arm that picked up samples of Martian soil and took them into the body of the probe for automatic experiments. Two experiments revealed what could have been signs of life: in one of them a sample of soil was mixed into a soup of nutrients that would have supported bacteria on earth and something in the soil decomposed the soup in the

115

*Viking I's meteorology
instrument which
relayed daily weather
reports of conditions
on Mars, after landing
on the surface of the
planet on 21 July
1975*

way that bacteria would. Then some samples of air brought from earth
reacted with the Martian soil as if it contained minute green plants that
could photosynthesise.

However, these observations were certainly not conclusive. There
were no other signs of life, and it is possible that the soil of Mars contains
catalysts that can bring about processes that closely resemble those that

116

are essential to life. If this is so, then there may have been similar chemicals on earth before life started, and the first forms of life may have superseded the catalysts and their activities.

It is obviously tempting to send a soft-lander to Mars to collect some samples of soil and return to earth with them. The Russians have done something similar on several occasions, the first in 1970 when they collected moon rock. There is, of course, the theoretical risk that we might import some new strain of bacteria, toxic and appallingly infective because no one on earth would have any resistance to it, but the recent development of genetic engineering means that the scientists need not take very elaborate precautions to avoid such a disastrous epidemic. Yet a soft-landing followed by a return journey would still pose the same problem as those of the Viking landers. Each Viking sampled only one spot of Mars: a return-journey mission would bring back soil from only one place. What is required is a roving tester. This is a tricky project, but not an impossibly difficult one: the Russians sent a 'rover' to the moon in November 1970. The Mars rover would need TV eyes so that the researchers on earth could monitor its movements, and it would need to be a moderately intelligent robot because of the length of time (19 minutes) it would take messages travelling even at the speed of light to pass to and from earth. The rover might achieve nothing, or only maintain the ambiguity that exists now about life on Mars, but it might on the other hand find life – the first to be discovered beyond earth. It might also find traces showing that life had evolved and then been wiped out as conditions on the planet changed.

JUPITER: THE STAR THAT DIDN'T HAPPEN

There is an enormous gap between Mars's orbit and that of Jupiter, which travels at 486 million miles from the sun. Though Jupiter is the largest planet in the solar system – it has a diameter of 289,250 miles – it is tiny by comparison with the distances between itself and its neighbours. Its density is small – 1·3 – but its mass is vast, two and a half times that of the total of all the other planets. The earth would fit comfortably into Jupiter's famous 'Red Spot', the very long-lived swirling storm that was first sighed in the seventeenth century.

Jupiter spins once every ten hours and this means that its outer atmosphere is vigorously stirred and swirled. It also explains Jupiter's shape – it is very recognisably flattened at the poles. The equator has a very distinctive bulge that is caused by the rapid rotation of the planet.

Jupiter has its own satellites, fourteen at the last count: these are not tiny balls of rock – four of them were seen by Galileo with his first

117

telescope in 1609, and two of these, Ganymede and Callisto, are as large as the planet Mercury. Their density, however, is low – not much more than the density of Jupiter itself: they must contain ice or condensed gases. The other two, Io and Europa, which lie nearer Jupiter, have densities near to that of the earth and are probably similar in structure. This pattern – the two dense satellites close to Jupiter and then, further out, two less dense ones – is similar to the pattern of the sun and its planets. Likewise, Jupiter might have qualified as a star. It seems that a large ball of the original cloudy stellar material collected to form the planet, but the temperature was too low to set off the nuclear reaction that would have made Jupiter a star instead of a planet. At its centre is a tiny dense, rocky core and this is surrounded by a thick shell of liquid hydrogen. This layer is so compressed that it has some of the properties of a metal – it will conduct electricity, for example – and is known as liquid metallic hydrogen. Outside this is another thick shell, this time of ordinary liquid hydrogen. This also contains about eleven per cent of helium and small amounts of methane, ammonia, water and carbon dioxide. If there were more of this mixture, it would be even more compressed by its own gravitational force and would generate an enormous amount of heat – Jupiter itself radiates more heat than it receives from the sun. With a larger mass of hydrogen, this gravitational heating would be enough to set off the hydrogen fusion that is the source of the energy emitted by the sun and by the other more distant stars. This internal heating helps to explain the electrical storms on Jupiter which are caused by the continual stirring of the atmosphere by gigantic convection current.

These storms were among the striking occurrences seen by the two US Voyager space probes as they passed Jupiter in 1979. Voyagers 1 and 2 left earth in 1977, directed towards Jupiter and Saturn and, in the case of Voyager 2, destined afterwards for Uranus. They were able to measure the speeds of the winds on Jupiter – a quite astonishing 340 mph – and also to take a close look at the Red Spot. It is difficult to believe that this is a storm, simply because it is so permanent – it has been in the same place since the planet was first seen. It must be sited over some unusual feature on the planet that generates vigorous vertical winds.

The Voyager spacecraft sent back startling pictures of the Galilean satellites of Jupiter, the most surprising of which came from Io: the observers actually saw a volcano erupting near the rim of the planet. Other pictures show that there are volcanoes on Io that are either still active or were recently so, and this probably explains why the surface of the satellite is so smooth. The volcanic eruptions have wiped out any traces of impacts from meteorites.

Europa, Io's neighbour, also shows no craters, but the reason for this seems entirely different. The close-up pictures from Voyager 2 show a whitish planet, criss-crossed with red lines. The whiteness suggests ice, and astronomers believe that Europa has an ice-coating more than 60 miles thick; the veins crossing it are probably fractures in the ice, possibly coloured by chemicals from below.

Jupiter's two outer Galilean satellites, Ganymede and Callisto, are slightly unusual: from their densities they must be largely composed of ice, yet both show impact craters – on Ganymede one can even see streaks of ice that must have been thrown up by the impacts of meteorites. Callisto, the outermost, is probably the oldest of the four satellites, and it shows its age by the density of the impact craters on its surface.

The Voyager spacecraft sent so much information back to earth that it will be a long while before it is all analysed and studied. Certainly it will yield a clearer picture of the nature of the Galilean satellites, and it will also help us to understand the astonishing meteorology of Júpiter and possibly of our own planet as well.

SATURN AND ITS RINGS OF ICE

Moving inwards from Uranus, the next planet is Saturn. The gap between the two orbits is enormous: while Uranus is 1793 million miles from the sun, Saturn is less than half that distance, at 890 million miles. It is one of the largest planets of the solar system: with a diameter of 75,000 miles, it is second only to Jupiter and nine and a half times bigger than the earth.

Saturn's appearance is familiar to millions who have never seen it. In the seventeenth century, soon after the telescope was invented, the astronomer Christian Hughens recognised that there was a ring around the planet. Until recently, it was believed to be the only planet with rings.

Ask anyone to sketch a planet, and the odds are that they will draw Saturn. Although other planets have rings, none can match the richness of Saturn's display. The rings are very conspicuous, but in fact they represent only a tiny fraction of Saturn's mass; they are probably composed of ice crystals, the smallest about a centimentre across, the largest the size of a football. In a telescope the rings seem to be smooth, but this is an illusion. Astronomers cannot explain their elaborate patterns with any certainty, even though they were first seen and identified by Galileo, more than 300 years ago.

Saturn is so distant – a minimum of more than 750 million miles from the earth – that telescopes can tell us little about it. The revelations of the Voyager probes therefore came as a great surprise. Voyager 1 flew past Saturn in November 1980, taking a close look at its largest moon,

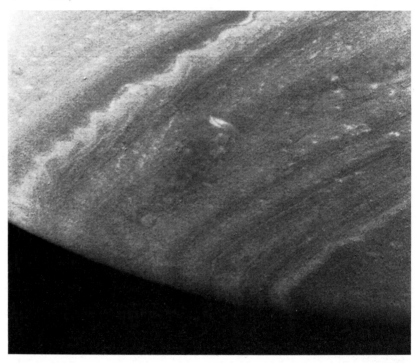

A Voyager 1 wide-angle image of Saturn's south polar region and mid-southern latitudes, showing cloud features separating here into light and dark bands of waves and eddies. This picture was taken 265,000 miles away from Saturn on 12 November 1980.

Titan. Voyager 2 flew past in August 1981. The most striking discovery of both probes was the detailed complexity of Saturn's rings.

Their structure is mysterious. There are recognisable rings with conspicuous gaps that seem permanent: the Cassini Division, for example, has been there for at least a century. It is difficult to understand why the rings are permanent. There must be collisions among these vast assemblies and, if this is the case, some particles should travel faster and others slower as a result. The faster ones would move away from the planet, and the slower ones drop towards it. Yet this doesn't seem to happen. The Voyager probes even saw what seemed to be rotating spokes within the rings, and a braided pattern in an outer one.

The explanation must be connected with Saturn's moons. These bodies orbiting the planet produce their own gravitational effects, and it must be these that keep the rings in existence. The pull of a moon must prevent the ice crystals from falling towards Saturn itself.

Saturn had fifteen moons at the last count, but the most striking of these is Titan. It is indeed the most striking moon in the solar system – the largest by far, with a diameter of 3437 miles, and the only one with a substantial atmosphere.

Voyager 1 was programmed to fly close to Titan, and it sent back revealing pictures and a wealth of information about Titan's atmosphere. The surface of the moon was invisible, even to the probe: Titan was covered with a mottled cloud, orange in colour, and there seemed to be a polar cap. It has been known for a long while that the density of Titan, 1·4, is greater than that of Saturn, which means that Titan has a small rocky core.

Its atmosphere contains mainly nitrogen and methane, but the action of the ultraviolet rays of the sun has converted some of the methane into

---

Saturn's satellite Dione taken by Voyager 1 on 9 November 1980 from a distance of 2.6 million miles. The bright spots may be rays emanating from impact craters on Dione's surface.

ethylene, acetylene and hydrogen cyanide, among other chemicals. Although the atmosphere contains no oxygen, it contains many of the compounds that must have been formed on earth in the processes that led to the evolution of life. On Titan the process probably stopped at this preliminary stage because the temperature on its surface is so low, $-180°C$, but if we could follow the reactions caused by sunlight there, it would tell us something about the start of life here. Titan is so very cold that the methane is partly solid, as well as a gas and liquid, and even some of the nitrogen will be liquid. Titan's rocky centre is dark under the clouds, with lakes of liquid nitrogen and methane, and occasional showers of other liquid organic chemicals. It is easy to see why life couldn't originate out there.

Saturn itself is even less hospitable. The planet that is seen at the centre of the rings is large – 75,000 miles in diameter – but the density is very low: 0·7, roughly the density of balsa wood. There is only a tiny rocky core, but in fact this is probably several times as large as the earth, and surrounded by liquid hydrogen that is so dense that it behaves as a metal. This in turn is surrounded by a layer of liquid hydrogen at a temperature near absolute zero ($-273°C$). There are traces of methane in Saturn's atmosphere, and even of ethylene and acetylene, chemicals that are precursors of life. But their presence shows only that the sun can cause chemical reactions: there is no possibility of life appearing on Saturn.

8

SKYLAB

By 1969 NASA had managed to put people on the moon and bring them back, and it wanted to see if man was suitable as a space settler. The technology of manned space exploration was well advanced, and this technology was both reliable and adapted to its task. It wasn't certain that man would be.

NASA therefore decided to launch a 'space house' called Skylab – vastly more roomy than the Gemini and Apollo space capsules – where astronauts would live for longish periods leading a 'normal' life.

The Russians had similar plans for rather different reasons. Once they had lost the moon race they announced that the building of space stations was far more valuable than landing on the moon. Though the statements were propaganda, the Soviet achievements were real: on 19 April 1971 they launched a space station, Salyut 1. They then notched up a first by sending men up to join the orbiting laboratory: Vladimir Shatolov, Aleksei Yeliseyev and Nikolai Rujavishnikov went up in Soyuz 10 and docked with the Salyut, but returned without getting into it. On 6 June 1971 Georgi Dobrovolsky, Viktor Patseyev and Vladislav Volkov went up, joined the Salyut and stayed there, using the combined Soyuz and Salyut as a home in space. The Russians had already perfected automatic docking and undocking, so that orbiting space stations could be supplied with food and oxygen. The Russian development of space stations started before Skylab, and it continued, very successfully, long afterwards: it is described in a later chapter.

The original plans for Skylab were that the astronauts would work for regular periods – say an eight-hour day – eat food that could imaginably be tolerable for months on end, sleep for a reasonable undisturbed period (unlike the earlier astronauts) and have time off for leisure – reading, listening to music and playing games. Although there were experiments to be performed, the aim would be a satisfactory sojourn in space, not the accumulation of a vast collection of scientific results. In the event, it didn't work out quite like that.

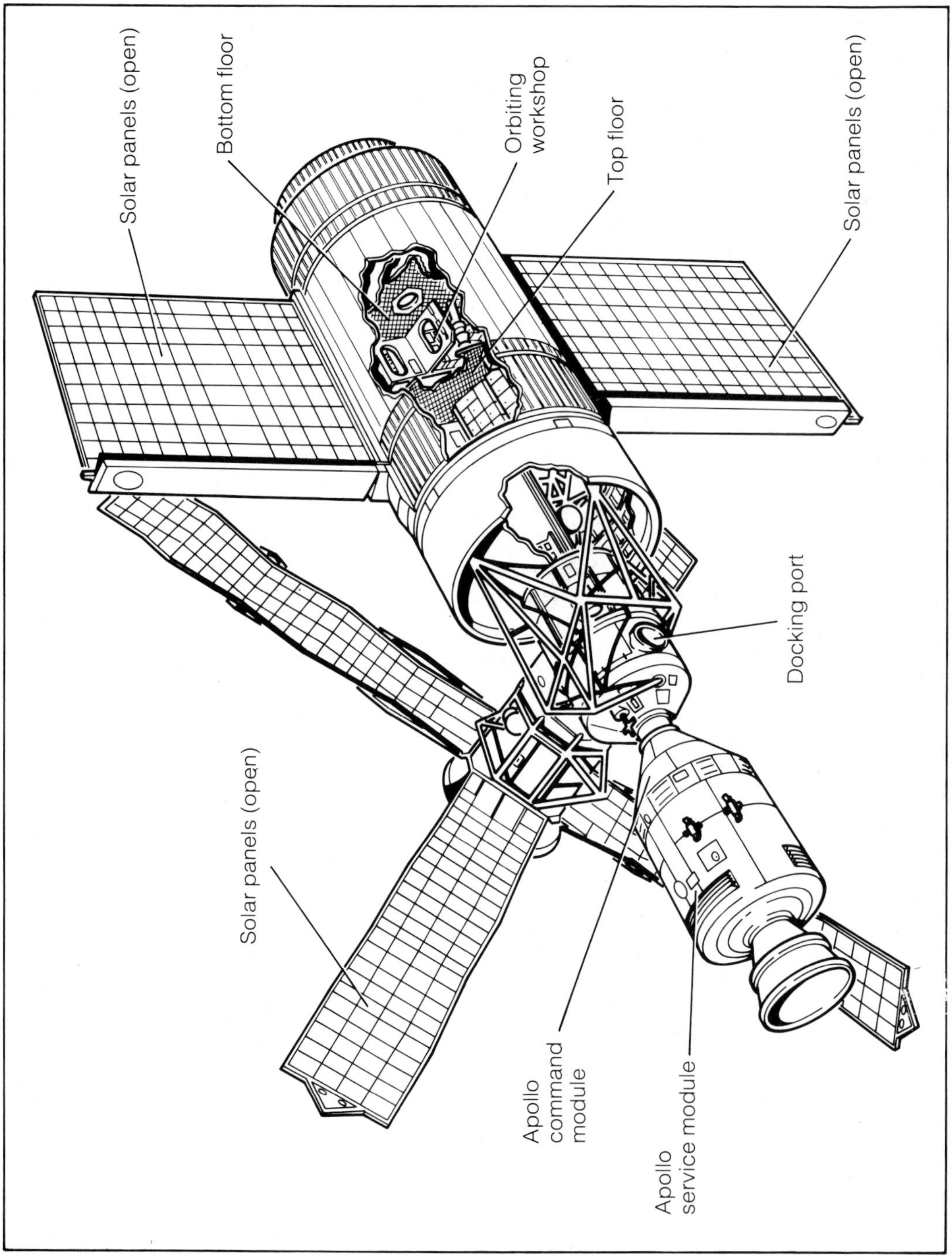

Solar panels (open)

Bottom floor

Orbiting workshop

Top floor

Solar panels (open)

Docking port

Solar panels (open)

Apollo command module

Apollo service module

Skylab

Skylab was originally designed during the period leading up to the moon landings, and the plan was simple. The astronauts would go up in the usual way, on top of a three-stage rocket, and the first two stages would fall away once they had burned out. Then the third stage would burn, but it would stay attached to the command module containing the astronauts. Once it had burned out it would be allowed to cool; the gases and liquids that remained from the combustion would be removed; and the astronauts would set up house among the now-useless rocket pipes and taps. This seemed a splendidly economical idea.

It was, however, somewhat improbable. It would be extremely difficult to put any accommodation into the third stage before the rocket was used, and tricky to do so in orbit; and there was no certainty that the exhaust gases could be cleared out well enough to make the atmosphere safe.

In mid 1969 the idea was modified. Skylab was to be constructed from the empty shell of the third stage of a Saturn rocket, but it would be built and fitted out on earth as an orbiting home and workshop and then taken into space on the first two stages. Once the home was in satisfactory orbit, the astronauts would be taken up to it on another Saturn rocket. This doubled the launching cost, but it was a more human approach.

The Skylab that was eventually launched in 1973 was a splendid piece of engineering, although it was rather alarmingly damaged during the launch. By space standards it was vast, with around 10,000 cubic feet of space inside – about the same amount as there is in a suburban house. It was differently arranged, but the peculiarities of living in space helped to make very good use of the space. For example, the astronauts could sleep in a vertical position – at right angles to the floor – because they were weightless and floating. You can't hang people up to sleep in a suburban house.

Skylab had two floors, separated by a metal grid, and was designed as if it were a cylindrical house, standing on the rocket's base. The upper part, about a third of the volume, was the experiment chamber, which was used to see how easy it was to move about in zero gravity. Moving, it turned out, was simple, and stopping impossible; but it was a good place for acrobatics. The lower floor contained a workshop, the sleeping area, the ward room, where some elaborate meals could be prepared, and the bathroom. Even by airline standards this was spacious, although the design was necessarily eccentric. There were suction pumps that drew waste and sewage into storage tanks – they wouldn't fall under gravity – and the shower was totally enclosed so that the water wouldn't simply drift weightlessly throughout the spacecraft. The whole set-up promised a rather civilised lifestyle.

Skylab's crew lived and worked in ordinary clothes rather than spacesuits. The uniform usually worn was a golden-brown two-piece: a long-sleeved T-shirt and elastic-waisted trousers, with elasticated wrists and ankles to prevent sleeves and trouser legs from floating up and down. In the colder parts of Skylab, such as the docking adaptor, the astronauts would put on jackets, and in the warmer parts they could unzip the lower part of the trouser legs. They found, incidentally, that they sometimes became surprisingly and uncomfortably warm when exercising because the heat that the exercise generated remained close to the exercising body: in space there is little convection to carry the heat away.

The astronauts' shoes were specially designed to prevent them from floating around the spacecraft: they had metal plates on the soles to lock into the metal of the gratings. There was a mushroom-shaped plate that gave a temporary grip, and a plate that fitted the slots in the grating exactly and fixed the astronauts very securely. They had to decide which they preferred.

The air in Skylab was a special mixture. After the disastrous launching-pad fire of 1967, pure oxygen wasn't considered suitable for a space home, so those in Skylab breathed a mixture of oxygen and nitrogen. This was much richer in oxygen than normal air and therefore at a low pressure – only a third of sea-level air pressure on earth. The low pressure contributed to the lack of convection, because the air that did circulate absorbed less heat than the normal, full-pressure mix. It also had a second, more disturbing, effect: the thin air of Skylab carried sound poorly, so the astronauts had difficulty in hearing one another. More worrying still, they had difficulty hearing the alarms attached to the receivers for messages from Mission Control. To ease both these problems there were loudspeakers scattered throughout the craft. Americans graphically call these 'squawkboxes'.

US astronauts in the Gemini and Apollo series had eaten special spacefood from tubes and squeeze packets, but the Skylab astronauts were able to eat more or less normal food, mainly dehydrated or deep-frozen. There were hot-plates and an oven for cooking, the latter being positioned near the roof of the ward room. The astronauts chose their own food, and opted for such extravagances as Lobster Newburg and Filet Mignon as well as the rather more mundane hamburgers. There was even a plan to take wine on the trip, but there were protests from temperance groups and NASA decided it would prefer to offend wine-lovers rather than the temperance lobby. Skylab carried soft drinks only.

Skylab was big enough to carry a 10-ton Apollo telescope mount, with a telescope and other instruments elaborately stabilised by gyroscopes, and there were plans to study the sun, X-ray stars, and a host of other astronomically fascinating objects. The astronauts also hoped to conduct some rather light-hearted biological experiments – pocket mice and vinegar gnats were carried to see if the odd day-length on Skylab affected their biological rhythms – and there were some preliminary investigations into potential space industries. NASA has always been under pressure to find economic justification for space research, and Skylab offered the first serious opportunity of seeing if any of the industries proposed as being particularly suitable for space projects would actually work.

The most frequently listed candidates for space industries involved either metallurgical or computer concerns: on earth, the computer industry has to take remarkable precautions to keep the atmosphere free from impurities. In space it is in theory possible to mix molten metals of very different densities to make totally novel alloys – the ingredients won't settle as there is no gravitational force – and also perfect and hence frictionless ball-bearings. On earth, gravity always distorts the beads of metal as they solidify to form balls, while in space, they should stay perfectly round.

Obviously Skylab's main experimental subject was to be the astronauts themselves. Astronauts had always deteriorated physically while in space, and this fact threatened to affect the design of settlements and of long-distance spacecraft. So Skylab pioneers were to measure almost everything they could about themselves to see what changes occurred – how much food they needed, how well they slept, what their concentration was like, and even how their moods came and went.

The start of Skylab's mission was inauspicious. It was launched on 14 May 1973, and was clearly in trouble within a couple of hours. The problem concerned the solar panels. Long stays in space require plenty of energy: the spacecraft has to maintain its own atmosphere and, in particular, it needs cooling as it spends much of its time in full, unclouded sunshine, well away from the shadow of the earth. Energy is also needed to maintain the life-support systems.

The ready availability of sunshine suggests one source of energy, and spacecraft are generally fitted with solar panels – assemblies of electronic devices that convert the energy of sunlight into electricity. For a long stay a large area of solar panels is needed; and in the case of Skylab this was too large to be fitted around its body before the craft was put into orbit. The plan was that the craft should take off with furled panels which would be opened out once Skylab was clear of the earth's atmosphere. There were two sets of panels. The panels in one set, once open, would be arranged in the form of a cross; the drawings of these give Skylab the appearance of a somewhat odd-looking helicopter. The helicopter blades were placed above the Apollo telescope mount and were designed to supply energy for this and for other experiments. In addition

127

there were two panels attached along the side of Skylab itself which, once unfurled, looked rather like short stubby wings.

A couple of hours after the launch it became clear that these panels were still furled, possibly because they had been damaged by an anti-meteorite screen that had floated away from the craft. This single failure put the whole project at risk. These panels should have been capable of supplying 21 kilowatts of electricity to the craft – an enormous amount for only three people, but crucial because they had to survive and work in space and have some energy in reserve. Without these panels the whole complex, vastly expensive Skylab project could have been doomed – a failure after only a couple of hours. However, the problem seemed soluble. The 'helicopter rotor' for the Apollo instrument mount could provide about as much energy as the wings and, if the astronauts were economical in their use of power the one source would be sufficient.

The loss of the anti-meteorite screen was also a problem, because it had been designed to act as a parasol. In bright sunlight the craft needed a shield to stop it from overheating. Without the shield the temperature inside the craft shot up to 65·6°C – far too hot for habitation.

For NASA it was vital that Skylab was successful. It had not attracted very much public interest and, as this diminished even further, so too did the interest of the US government that was providing NASA's money.

There was a real possibility that the craft would get so hot that the

Skylab 1 in orbit, but with only one set of solar panels unfurled. The failure of these panels and subsequent lack of energy put the whole project at risk.

128

aluminium body and the instruments inside would be at risk. The space scientists on the ground solved this problem temporarily by putting Skylab at an angle to the sun, but this of course meant that the solar cells didn't work fully. There had to be another way of keeping the craft cool.

The proposed solution was attractively simple. It was necessary to keep the sun off Skylab. Why not, therefore, put a sunshield up in space? The three astronauts who were to make up the first crew – Charles Conrad, Joseph Kerwin and Paul Weitz – would carry a roller sheet of aluminium-coated plastic in their spacecraft and fix it to the hull. It would involve working outside both Skylab and the Apollo spacecraft that was the ferry for the crew, but it seemed practicable. The crew rehearsed the manoeuvre on a replica of the craft in Houston.

Whatever was to be done had to be done quickly. Though the re-alignment of Skylab had brought its temperature down, the instruments aboard showed that it was filling with poisonous gases. These could be pumped out by remote control, but the worrying fact was that they came from plastics inside Skylab that were decomposing in the heat. Some, no doubt, were merely trivial bits of furnishing, but it was clear that electrical insulation was also being destroyed. If Skylab wasn't safe-guarded quickly, it would become a useless hulk.

Though the situation was desperate, the solution had to be sane. NASA recognised that it would be very risky to send astronauts up to spread the shield. They had very little experience in space-working, and this would be an extremely awkward task.

A simpler plan was evolved. The astronauts were to dock with Skylab and then spread either a folding parasol or an inflated one, while still maintaining contact with either the Apollo craft or the docking adaptor. They would still have to work in space, but they would be safely secured while they did so.

This approach was safer, but no one knew if it would be possible. The first stages certainly seemed as doomed as the launch itself. The astro-nauts arrived at Skylab and tried to dock. There was a docking probe on the Apollo craft to engage with the docking adaptor on Skylab and, after that, the two should have linked automatically. But the probe failed to engage four times in a row, and eventually the astronauts had to make the coupling themselves, wearing spacesuits. This said a lot for the ability of men in space to cope with problems, but not very much about the design of Skylab.

Conrad, the mission commander, tested to see if battery leaks or chemical processes had formed poisonous gases inside Skylab and found none, so Weitz went in to start work. Though the temperature of the main body of the spacecraft was at more than 100°F, the docking adaptor, where they would work on the parasol, was at a bearable temperature of about 50°F. It looked as if there would be no need for the special water-cooled spacesuits which had been brought along.

The first problem was to erect the parasol. The aim was stunningly simple: the parasol had to be poked out into space through an airlock in the docking adaptor. The adaptor was narrow, so there was a series of interlocking poles—like those on a sweep's brush - to push the parasol out. Having been opened, it finally had to be brought close to the hull of Skylab, then the interlocking poles were to be removed before the parasol was fixed in place. Somewhat surprisingly, considering the problems so far, this all went according to plan.

The three astronauts could now settle back to see how much damage there was, and what the prospects were for continuing the mission. Without the solar wings there could only be one, month-long mission. The rest of the programme, using the other two teams of astronauts, would have to be cancelled unless the solar wings could be repaired.

The problem was a simple one. Part of the meteorite shield that had broken away was wrapped around one wing, trapping it like a butterfly half out of its cocoon. To free it, Captain Conrad went out in a spacesuit, worked his way along the wing, and cut away the strip of metal that held it with a pair of metal-cutters. The wing moved outwards a couple of feet and then jammed because the hinges had frozen. Conrad worked his way along it and swung so that his mass helped it open. It moved out and immediately started generating electricity. Skylab was saved.

Naturally the technical problems weren't over, but the crew was able to deal with the rest of them. The final problem was the failure of part of the refrigeration system, just before the crew shut Skylab down and climbed back into the Apollo craft to return to earth. They made a perfect splashdown on 22 June 1973 and were duly taken aboard the aircraft carrier *Ticonderoga*. They had been in orbit for twenty-eight days, longer than ever before, and while it would take time to check their physical condition there was clearly no problem so severe that longer trips had to be ruled out. NASA began to make plans for a two-month trip.

The second crew - Alan Bean, Owen Garriott and Jack Lousma - joined Skylab on 26 June. Technical troubles started after only a few days. The technicians at NASA detected a leak in the fuel tank of one of the manoeuvring jets of the Apollo service module and the engine was shut down. As this module served as the ferry to and from Skylab, if it were to fail the spacemen would be in danger of being stranded in space. Fortunately, though, the module had four manoeuvring jets, and the loss of one of them wasn't regarded as crucial.

Then, on 1 August the engineers at Houston discovered that a fuel tank on the same vehicle was too cold. Heaters were switched on, but unfortunately this revealed a leak so a second engine was shut down. Now the problem was serious. Without two of its engines the Apollo craft was manoeuvrable only with difficulty and there was a risk that the rather exotic fuel might leak into the engine and damage it. If this happened, Apollo was doomed. NASA started to think of bringing the astronauts down before their ferry became useless.

NASA quickly realised that this simple solution – a space equivalent of taking to the lifeboats – was a poor idea. If the Apollo module was going to deteriorate, it might do so en route. A better possibility would be to see what happened to the craft if the astronauts were left in orbit, where they had sufficient supplies for a very long stay, and prepare to rescue them if the situation worsened.

Were it to become advisable to launch a space rescue, it would be the first one in history. As it happened, this was a time when it would be possible, because there was another Apollo and its Saturn rocket being prepared for the third Skylab crew. If necessary, preparations could be accelerated and the craft sent up with only a couple of crew to collect those stranded in Skylab. It would be expensive: it would mean that the third, longest Skylab trip would have to be cancelled, or to go ahead using yet another Apollo – but it would be possible. And, while NASA waited to see if the rescue would be needed, the Skylab 2 crew could start their repair work.

Their first task was to fit an improved sunshade. The parasol was still in position and giving shade, but Skylab needed a larger shield. This was duly erected. Meanwhile, back on earth, Houston's engineers had decided that the module had sufficient engine power to be able to guide itself on its return flight. Furthermore, there seemed to be no danger of contamination. The crew proceeded with its research. While the first Skylab trip had been only slightly longer than the longest Russian trip, this double-length voyage might provide interesting results to experiments monitoring the effects of space upon the human body. When this crew splashed down on 25 September they were to receive the very close attention of the space medical teams.

Before that, however, they would need the attention of the pick-up team. Without its full manoeuvring capability, the craft had to be directed back to earth rather unconventionally, and eventually it landed upside down in the sea. The craft was righted and the crew taken off. It quickly became clear that they had not been disastrously weakened by their record stay in space, and NASA prepared to send the third crew up. In fact, the doctors recognised that the second crew had benefited from the experience of the first: because they had conscientiously kept to a rigorous exercise schedule they suffered fewer ill effects, even though their stay had been twice as long.

Skylab 2 astronauts leave their spacecraft accompanied by a NASA medical team after they crashed down having spent 28 days in space

The third flight took off on the morning of 16 November and docked with Skylab that night. The crew – Gerald Carr, William Pogue and Edward Gibson – were all making their first space trip and would be up, if all went well, for an even longer time – eighty-four days. It would be a more severe test still of the possibility of living in space. It would also give the researchers the first chance ever of looking at a comet from space. Comet Kohoutek would be bright in the sky towards the end of the year, and astronomers were keen to observe it without having to look through the distortions produced by the earth's atmosphere. The results of those observations are still being analysed.

The third crew splashed down successfully on 8 February 1974. They left Skylab ready for visits by other astronauts, but nobody really thought that there would be a further trip. The cost of the visits was enormous – around four million dollars a week – and NASA wanted some time to determine the nature of the next experiments, assuming that more money would be available.

Even empty, Skylab could present problems. When the craft was abandoned NASA reckoned that it would stay in orbit until at least 1983, possibly 1986. As there were plans to put up a space shuttle in 1979, there was a chance of taking a new crew up to Skylab and using it again. But right from the start there had been doubts about the life of the spacecraft. In Britain, scientists at the Royal Aeronautical Establishment (RAE) calculated that Skylab would come down in May 1979. The RAE claimed that NASA hadn't allowed for the effects of an unusually large number of sunspots which appeared at that time and which warmed the earth's atmosphere so that it expanded and therefore grew up towards Skylab's orbit. If the spacecraft ran into a denser atmosphere than NASA had predicted, it would be slowed down and would start to fall towards the earth.

The RAE, it turned out, was right, or was on the right track. In February 1978 NASA recognised that Skylab was circling in a lower orbit than intended. By the end of the year it would be so low that it might re-enter the atmosphere and return to earth soon afterwards.

This would be no trivial event. Skylab weighed some 85 tons and, although it would break up when it hit the atmosphere, even the fragments would be substantial. Most of the earth's surface is empty, of course – about seventy per cent is covered by water – but, because the spacecraft's orbit was between 50° north of the equator and 50° south, it would crash between those latitudes, and that area included many of the most densely populated cities in the world.

Bright in NASA's mind was the memory of the effects of the crashing of an unmanned Soviet satellite, Cosmos 454. This came down in Canada and scattered itself and its radioactive fuel over deserted country. The public was shocked to discover that such an occurrence was possible, and NASA was anxious not to be associated with threats to people on earth, even though Skylab contained no radioactive material. While they aimed to restore Skylab to its original orbit, or even a higher one, that was an ambitious and complex project. The first task was to make sure that it didn't come down any more quickly than necessary.

By early 1978 Skylab was orbiting about 214 miles up, with its base facing the earth and spinning slowly around its long axis. This gave it the maximum air resistance possible. NASA decided to turn it so that it was travelling horizontally, a more streamlined posture. It was time to waken the dormant mechanisms.

First, NASA sent up signals that were to recharge the batteries. This was complicated: because Skylab was spinning, the solar cells on the wing and helicopter vane faced the sunshine for only short periods at a time. The solar cells were to charge the batteries which were to supply energy to the computer for the control system that would use nitrogen-gas thrusters to correct Skylab's flight, and for the big gyroscopes that helped to stabilise Skylab and to give the control system some reference. These systems depend on each other. Without electrical power the

133

computer and control thrusters wouldn't work, but they had to work if NASA was to stop Skylab from spinning. And Skylab had to be stopped from spinning if the solar cells were to operate fully to charge the batteries. Further, the whole assembly had been shut down for five years since the departure of the last crew.

Gradually, over a period of months, NASA worked through a strategy that would bring the spacecraft under control. Some electricity was used to drive the control system, and that improved the yield from the solar cells. By June the spinning was under control and the solar cells were working fully. The control mechanism was switched on and, apart from the one gyroscope that had failed while the third crew was aboard, everything functioned correctly. Skylab was turned to travel end-on, with its long axis parallel to the ground, so that air resistance would be as small as possible.

All seemed well. Then the batteries cut out, the control system ceased to work properly, and Skylab started rolling. It turned out that the

View of Skylab space station cluster. This photograph was taken by the Skylab 3 crew who left Kennedy Space Center on 28 July, returning on 25 September 1973.

134

batteries were overheating. They heated up as they were charged, and the plan was that the controllers on the ground should switch the charging over to another set of batteries when one set overheated. But there were only three control stations and for long periods there was no contact between the ground and Skylab. The battery charging couldn't be changed over frequently enough.

Once NASA had recognised the problem, they worked out a way of covering the long gaps in communication. An extra ground station was set up so that the gaps became shorter. Skylab was once again travelling under control and NASA began to plan a rescue involving the Space Shuttle, which was being designed and built at this time.

The Space Shuttle was to be a space bus that would take people and equipment into space relatively cheaply. It would be used for working in space, as well as for experiments, and it would have a fairly spacious cargo hold to carry satellites for orbit or repair, or scientific instruments for research. In order to rescue Skylab – which actually meant increasing its speed and putting it into an orbit further away from the earth – the Shuttle would be used to carry the Teleoperator Retrieval System (TRS), a radio-controlled rocket system, costing $20 million. The Shuttle would carry it up and rendezvous with Skylab. Still under radio control, it would be fired to take Skylab 75 miles up to a safe orbit 225 nautical miles above the earth.

Alas, none of this happened. It would have been an expensive project, although the value of Skylab would have justified it. But the Shuttle wasn't ready in time and there was no other way of getting the TRS to Skylab. NASA prepared for the spacecraft's re-arrival, in large pieces, on earth. This was anticipated for the evening of 11 July (BST), and newspapers carried maps showing the area at risk – a band around the world extending from 50°N to 50°S. Skylab would be travelling at 17,000 mph when it struck, though nobody knew exactly when this would be.

The £1160 million spacecraft sent out its last radio signal at 5·11 p.m. BST on its 34,981st orbit and broke up over the South Pacific and Indian Oceans. Most of the debris fell into these waters, although some pieces arrived with a sonic boom and a blaze of light over Western Australia. They passed over Perth and over a small town, Esperance, to the south, and most of it fell in a desert near Kalgoorlie. Houses were shaken, but there seems to have been neither serious damage nor serious injuries. From beginning to end, Skylab had posed problems but none of the possible disasters had occurred. The flight had been a qualified success.

On the one hand, Skylab proved to be a demonstration that the most beautifully planned and carefully executed technology can fail and on the other that, although there is much that can be achieved with unmanned equipment, sending men into space is sometimes the only way to perfect space devices and thereby save large amounts of money. Skylab was also the first ever experiment on living in space. Astronauts had experienced weightlessness in earlier capsules, but they had no real opportunity to, say, float from place to place in a spacecraft. They hadn't tried cooking and eating ordinary food in space, and they hadn't tried working there in the way that settlers would eventually want to. Skylab gave some insight into the pleasures, and the quite remarkable frustrations, of such a lifestyle.

When designing and launching Skylab NASA had always had space settlements in mind, although this aim sometimes became obscured by a frenzied determination to get immediate research results that would justify the money spent on the trip. But space was originally seen as territory to be developed in a way that paralleled the development of the American West. It was to be investigated by explorers – the moon landing had been typical – and then by working groups. Finally young families would move into space, as they had done into the West. NASA quickly developed plans for space settlements of ten thousand people or more.

It all depended, of course, on whether people could actually cope with the problems of living in space. The evidence to date was disquieting, if not actually damning, but it all tended to show that a weightless existence harmed the body. Some of the damage was temporary: for example, muscles that weren't used got weaker, but they quickly got stronger again when the astronauts returned to earth, and the weakness didn't matter in space. However, there seemed to be various other effects that might prevent any potential settlers from living and working out in space for an unlimited period of time.

It would have been surprising if the zero gravity in space did not have some effect upon the human body, in fact it had been known for some time that there were obvious and predictable alterations. Normally gravity pulls blood and other liquids downwards. Blood, in particular, tends to accumulate in the legs, and the heart and blood vessels are responsible for pumping it uphill again. In space the body fluids accumulate in the upper part of the body, because the mechanism that compensates for gravity continues to operate even when there is no gravity to operate against. Thus the astronauts had distended, flabby, florid faces and they never felt really hungry, although when they ate they too quickly felt satiated. Indeed the experience of earlier, more confined, astronauts led NASA to a significant underestimation of the amount of food the later astronauts would need. The Skylab crews ate far more than their predecessors, especially after using a device that reversed this 'floating liquid' effect.

136

In a real space settlement a gravitational force would be useful – it is easier to cook for example, if the food doesn't float out of the saucepan. A gravitational force might in fact be essential to life, in which case it could be simulated by rotating the settlement. This would give the feeling of a gravity-like force pulling the occupants outwards. The designers of Skylab wanted to see if life was possible without such complexities, but they did propose some simple gadgets which would enable the astronauts to avoid the physiological harm caused by living under zero gravity. Two particularly simple gadgets were tried – the lower body negative pressure experiment and a bicycle ergometer, a rather sophisticated version of the exercise bicycles sometimes used by flabby city dwellers who hope to reverse the effects of *that* environment.

The lower body negative pressure experiment is essentially an aluminium barrel that the astronaut can lower himself into and then seal off at waist level. The air in the barrel can be pumped out, thus creating a suction that expands the blood vessels of the lower part of the body and draws blood down into them. This obviously reduces, at least temporarily, the discomfort caused by the liquids that accumulate in the upper parts of the astronauts' bodies, and gives the body's blood-return mechanism something to do.

After landing the astronauts were, of course, suddenly subject to gravity and the blood tended to pool in the lower parts of their bodies, sometimes causing them to faint. Traditionally they were expected to stride along a carpet on the deck of the aircraft carrier that was their first home after landing. They wore inflatable rubber trousers to help offset the problem, but the reduced pressure gadget gave even the Skylab crew, the longest space residents to date, a sporting chance of reaching the end of the carpet. The main drawback of the gadget was that it was so obviously a rather simple-minded attack on what was a genuine problem that the astronauts felt unhappy using it. Exposing part of the body to suction was a bit unnerving – Kerwin, a doctor and one of the first Skylab crew, felt that he was about to be sucked into the machine. Having neglected to use it regularly, on the thirteenth day of his trip he climbed into the machine and found that his blood vessels, now somewhat weak, expanded dramatically. He felt queasy and dizzy and hurriedly got out. He was the astronaut who came nearest to fainting when he walked along the carpet on the deck of the aircraft carrier soon after splashdown: not only had he neglected his exercise routine, he had also failed to inflate his rubber trousers properly.

On the whole the astronauts were very concerned about their physical condition. They were normally very fit, active people, and had no intention of going to seed in space. The bicycle ergometer became popular, but there were problems with this as well. The astronauts had to secure themselves while using the bicycle: they wedged themselves with cushions against the ceiling, pulled down on the handlebars, and locked their shoes on to the pedals. Then they found that their space uniforms weren't suitable athletic gear, so they became sweaty and smelly. They could clean themselves up, but this was a lengthy process using a damp cloth. They all persevered, however, and found that they looked forward to their rides. There is nothing so physically undemanding as pottering about, weightless, working on experiments, and the astronauts all wanted to spend long periods of exercise toning up their systems. They felt and looked better afterwards.

Despite using the exercise bicycle and the reduced pressure device, the astronauts still deteriorated physically. They were probably the most meticulously monitored people in history: the medical teams back on earth knew everything about their muscles, their lungs, their body fluids and their hearts. If they continued to deteriorate, then a zero-gravity spacecraft residence wouldn't work.

The second, two-month Skylab trip provided a medical revelation. Somewhere between the thirtieth and fortieth days the deterioration stopped and the astronauts even began to get fitter. The number of red cells in their blood, which had fallen steadily during the first fortnight or so, began to rise again.

On the third trip the number eventually returned to normal: as red blood cells carry oxygen, this was important. Equally important was the change in the behaviour of what are called blood electrolytes. These chemicals are involved in the action of muscles and, while one might argue that it doesn't matter if, say, the arm muscles of astronauts deteriorate, one can't say the same about heart muscles. There was a risk that the loss of electrolytes would lead to an irregular and risky heartbeat, but the change reversed itself before any damage was done.

One change seemed continuous: the bones of the astronauts lost calcium and became weaker. This was never dramatic, but it was worrying. It could be that enough exercise would maintain the amount of calcium but it might also be that an astronaut returning after a very long trip would require some medical treatment before he could move about in normal gravity.

The medical supervisors faced a dilemma: if the astronauts continued to deteriorate during a space stay, then they would be worried about leaving them up there; but if, on the other hand, they adapted, which seemed likely, they might have difficulties re-adapting to conditions on earth. However, on the whole the results of the medical experiments boded well for space settlements, and so did the actual experience of working in space, though Skylab highlighted the pitfalls.

The problems caused by living in a low-pressure atmosphere – difficulty in hearing and talking – also extended to eating: the smell of the food didn't reach the astronauts and it sometimes seemed tasteless. As mentioned before, the methods of cooking and eating on Skylab represented one of the attempts to make living in space as normal as possible. On earlier space trips the astronauts had eaten from tubes or bags and simply squeezed the food into their mouths. On Skylab there were chairs and a table, knives, forks and spoons. The attempt at normality didn't always work. The chairs, which comprised a couple of bars so that the astronauts could lock themselves into position, were neatly adjusted to the height of the table. But in space you have to use muscles to stay in a sitting position, and this was fatiguing: the astronauts therefore stood to eat, their feet locked into a grid on the floor. This, of course, meant that the table was too low. And although in popular accounts of life in space you don't need tables, because your dish of food simply floats to where you want it, you can hardly expect a dish of cornflakes to stay in position while you spoon them up.

The astronauts prepared their breakfast trays overnight, putting cans of, say, sausage and corned beef hash into holes in the tray so that everything could be heated in an oven in the morning. When it was breakfast-time they would get out their knives and forks, open the cans, and settle down to a meal. Or that was the idea. The reality was more trying. When an astronaut opened the box containing the knives and forks, the implements started to drift around the cabin. Eventually they devised ways of fixing knives and forks down with elastic bands, but they had the same problem with the canned food. The lids had to be taken off, and there was a good chance that the breakfast itself would drift around the cabin, accompanied by any unopened tins – these weren't a tight enough fit in the holes they rested in. Furthermore, the ovens didn't really heat the food efficiently so, as well as being bland, it was always tepid. Because the astronauts' faces were suffused with blood their appetites were poor. They had chosen their own menus before leaving earth, but the food was considerably less appetising when prepared in space. The first crew tried adding flavour to their food with salt and pepper, but experienced trouble with the salt- and pepper-pots. The third crew finally solved the blandness problem by taking some powerful spices with them.

Working while weightless proved endlessly irritating. Everything in space stays, as Newton said it would, just where it is until disturbed by a force. The problem was that it was impossible to avoid disturbing anything that was loose. When the astronauts opened a drawer, everything inside cascaded out. With hindsight, it seems surprising that some of the experimental equipment was stored in chests of drawers. When an astronaut pulled out the top drawer, he had to contend with its entire

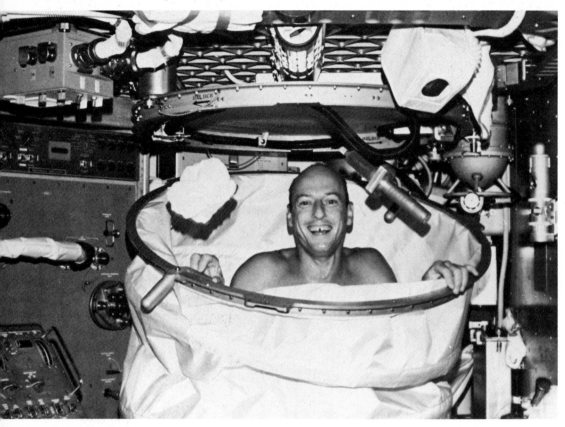

Skylab 2 commander, Charles Conrad, taking a hot bath aboard Skylab

FACING PAGE
above, *Tom Stafford (US) and Alexei Leonov (USSR) at the hatchway of the Apollo docking module and the Soyuz orbital module during the Apollo-Soyuz link-up*

below, *Soviet orbiting station consisting of two Soyuz spaceships*

contents; then, when he wanted to replace the equipment he had to deal with what should have been in a lower drawer but had drifted up to jam the upper one.

Carrying out even simple tasks was tricky. For example, Carr and Pogue, two of the astronauts on the third trip, had to repair a camera designed for ultra-violet photography in space. The first part of the sequence involved consulting the workshop manual – a spring-bound loose-leaf book that on this occasion snapped open, allowing the pages to drift off. (One of the design proposals for future stays in space is an aerodynamic table, where paper will be held in place by air pressure.) Then they had to get the tools they needed from a toolbox, which produced a cascade of hammers, spanners and screwdrivers. They had to replace those they didn't want and put the rest into pockets in their overalls, trying to secure the pocket flaps so that the tools didn't drift out again. Using the tools presented further problems. They weren't specially designed space tools, so the astronaut had to fix himself into

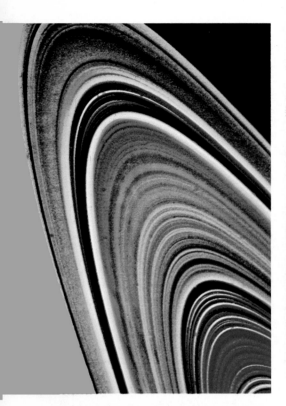

position before using, say, a spanner, which would otherwise cause him to rotate around the nut he was working on, or a hammer, which would send him spinning across the workshop. Then there was the problem of securing the components of the camera as it was dismantled. The astronauts simply had to learn to keep making a round of their floating clouds of tools and parts, shepherding them into a compact mass. The difficulties were compounded by the fact that they had no experience of looking for floating screws and springs, and there were times when they simply couldn't see them. It was only a small consolation to them to know that everything ultimately arrived in one place. The air-conditioning system of Skylab circulated the air in through one mesh screen and out through another. Anything that was not secured was eventually carried to the exit mesh.

These unpredicted problems meant that every task took much longer than the space planners had anticipated. And as the Skylab trip continued, Mission Control tended to add extra tasks, so the astronauts came to lead rather harried lives. These extra tasks represented a departure from the original idea behind Skylab: to test living conditions in space. But spurred on by its fight for funds, NASA felt bound to produce results that would show, or at least point the way to, economically worthwhile results.

The extra demands made on the astronauts led to history's first ever strike in space. One day, at the end of the sixth week in orbit, the third crew simply stopped taking orders from the ground and did what they wanted to. They took pictures of the earth and the sun; they looked through the telescopes; they tried odd little experiments in weightlessness. Throughout the trip they had been the crew most critical of the way that the equipment on Skylab worked or, as they often found, didn't work. Now they simply said that they had been harried enough. Mission Control quickly lightened their burden.

Living in space proved to be perfectly practicable. The difficulties with both the food and the drifting clouds of tools and experimental equipment could be solved as a result of the experience. As for the astronauts, they moved around by drifting through space, though they didn't always float by the shortest routes. They tended to travel vertically as anything else was very confusing, but they liked the aerial gymnastics and also the jet-propelled manoeuvring units that they could use as space scooters.

The scientific research – the observations of sun and earth – produced satisfactory results and demonstrated that for some kinds of observation there is no substitute for man in space. The industrial research was less encouraging. Gibson, for example, found that he could spin what seemed to be perfect lenses from a bubble of water in space. NASA rather hoped

FACING PAGE
above left, Voyager 2 image of Saturn's rings

above right, Voyager 2 view of Saturn

below, Voyager 2 picture of Jupiter's characteristic Great Red Spot – an area of 15,000 miles from top to bottom that the astronauts would find some way of freezing these water lenses and bringing them back to earth for study, but this didn't happen. The ball-bearing experiments didn't work either. Though it was easy to melt the metal, there was no way of getting it out of the crucible without producing a tail rather like that of a tadpole. A ball-bearing with a tail was a lot less useful than those made on earth. An ingenious solution to this problem might be found on future flights. The molten metal might, perhaps, be squeezed out to travel through a long enough distance to coalesce as a perfect ball. The problems that Skylab discovered were annoying but not disastrous, although industrialists on earth have been heard to say that NASA hasn't yet asked them if they actually want a perfect ball-bearing.

Another possibility was the creation of a space crystal industry. For micro-electronics and for ultra-strong materials there is a need for large crystals grown under careful control. Micro-electronics – the electronic chip is an obvious example – involves circuits etched on a single crystal, usually of silicon: long crystals grown as 'whiskers' have enormous strength and can be used to reinforce plastics. It might be possible to grow such crystals in space, although it isn't by any means certain that this would be the most economical method. It has not yet been established which useful products might be made in factories in the space settlements. But we do know from Skylab that there seem to be no physiological problems that would prevent people from settling out there.

9

THE RUSSIANS FIGHT BACK

The Americans may have won the race to the moon, but according to the Russians there never was a race. They certainly didn't have a rocket to match the Saturn V, but they were working on one. However, even with a powerful booster they wouldn't have put men on the moon. Their spacecraft weren't capable of it. The basic simplicity, not to say crudity, of the Russian space effort was revealed during the Apollo-Soyuz test project.

This was, at least in part, politically inspired. President Nixon was trying to soften the Cold War; Prime Minister Kosygin of the Soviet Union was in agreement. The magical word of the time – the few years following 1970 – was *détente*. The two superpowers would agree to mutual respect; each would remain happy with its present possessions and both would work for peace.

Manned space research was an obvious area for co-operation. Until this time the two countries had been operating as rivals. Had one spacecraft run into trouble, a spacecraft from the rival nation would not have been able to do anything to help, mainly because they used different systems for docking. It was obviously an odd approach to exploration and, once the moon race was over, co-operation became acceptable.

In May 1972 Nixon and Kosygin signed an agreement for a joint launch and link-up. In retrospect we can see that this suited NASA, who had one Apollo rocket and one spacecraft available, but no specific plans for them. It also suited the Soviet Union, who had run into a bad patch in their space flights and welcomed some help. No doubt they hoped that they would learn something about designing sophisticated spacecraft.

There had never been any great secrecy about the US space effort and there was none now. The Americans welcomed the Russians and taught them about Apollo. In turn, the American astronauts went to Baikonur and learnt about the Russian launchers and spacecraft. They were startled by what they found. Firstly they met Soviet astronauts and

143

heard some fairly true accounts of what the Russians had done during their space programme. Secondly they had the first foreign look ever at a Soyuz spacecraft. It turned out to be a simple, unsophisticated vehicle, little more than an unmanned spacecraft converted so that people could live in it. And the astronauts had hardly any control. The Russians had chosen the simplest possible system for controlling the craft: advanced modern technology had barely been used. While the system had worked fairly well in the past, it had reached its limits. For instance, the spacecraft was not controlled as it returned through the atmosphere. It simply spun: this kept its track accurate, but tended to cause travel nausea. Then there was no automatic inertial system for keeping the craft in the correct attitude. The astronauts had to look out through a periscope and fire trimming jets to align their craft. This was impossible if they were in the earth's shadow. Equally, the Soyuz had no on-board computer. On the Apollo, controls by the crew were channelled through a computer that converted these commands into action by the various thruster jets and monitored the movements of the spacecraft. The astronauts could issue whatever orders were necessary and, in emergencies, even use procedures that the designers had never envisaged. The freedom to invent such procedures saved Apollo 13 and its crew. The instructions for the Soyuz were carried on a roll of punched cardboard, and the crew was free only to choose which punched procedure to use. When they had to burn a rocket for a manoeuvre they timed it with a stopwatch, then waited for the ground station to tell them if the manoeuvre had been successful.

The technical problems to be solved before the proposed link-up could take place were formidable. The difference between the docking systems on the two craft was much more than just a matter of size, and both would need alteration. Because the Apollo spacecraft was potentially a moon-trip vehicle it carried more fuel; it was therefore decided that in this first ever international link-up, the Apollo should chase the Soyuz. Though the Apollo was in general more technically sophisticated than the Soyuz, this didn't apply to the docking techniques. The Russian system could dock automatically; the Americans did it by hand and eye. A minor problem was that the Soyuz was dull-coloured, which made it difficult to spot from the Apollo. It couldn't simply be repainted white because that would change the temperature inside. A compromise – part-white, part-green – was reached. The Soyuz was also fitted with beacons and flashing lights.

It was both expensive and difficult to alter the dimensions and mechanism of the actual docking adaptors, but without the alterations the two systems couldn't connect. Ideally, both a new Apollo and a new Soyuz system would have to be designed, built and tested on the ground and

left and below, *The Soviet Soyuz spacecraft as seen in earth orbit from the American Apollo spacecraft just prior to docking, 17 July 1975*

then in space before they could be used. This was unacceptable, so the Americans simply built an adaptor. As it was intended as a rescue system it had to be suitable for either craft.

The differences between the two craft were more than mechanical. The Russians breathed air, the Americans oxygen. The atmosphere in the Soyuz was more or less normal air at two-thirds the atmospheric pressure on earth; the Americans had pure oxygen at only a little above half the Soviet pressure. If a Russian moved straight into the Apollo he would get the diver's affliction known as the bends. In the end the Russians agreed to drop the pressure for long enough before transfer to obviate this problem.

The design problems were solved, the training finished, and on 15 July 1975 the Soyuz, with Alexei Leonov and Valery Kubasov, took off – for the first time ever in Russia, the launch was televised live. Seven and a half hours later the Apollo - with Tom Stafford, Deke Slayton and Vance Brand - set off in pursuit, in what was essentially an ever-increasing spiral. On 17 July, fifty-one hours and forty-nine minutes after the Soyuz had taken off, the Apollo docked, achieving the first international space link-up.

The crews moved between the two spacecraft, shook hands, shared meals and toasted each other with tubes of soup that had vodka over-printed on the label. The link-up was a complete success. In due course the craft separated. The Soyuz landed first, perfectly successfully. The Apollo followed, narrowly escaping a disaster caused by a poisonous gas that had leaked into it. But there was no accident. Technically and socially the Apollo-Soyuz test project had been a success. But, by then, the spirit of international co-operation that had prompted the flight had waned.

There were no further Apollo-Soyuz flights – there couldn't be, because there were no Apollo craft left and NASA couldn't raise any money to build more. To NASA, the value of the trip was firstly that the Americans found out just what the Soviet spacecraft and spacemen were like and, secondly, that the enterprise permitted it to launch the last Apollo and make use of the last discardable US rocket to put people into orbit.

The Russians gained an insight into the vastly more sophisticated spacecraft of the Americans, and at first it seemed that this might encourage them into direct competition with the Americans. However, the Russians were still having difficulty making a reliable rocket that was as powerful as the Saturn V. And, although they learnt about the design of the Apollo spacecraft, that was, after all, about ten years old. The Apollo-Soyuz joint exercise was expensive and mildly rewarding, but its main value was in expressing the rather temporary spirit of East-West *détente*. It was, too, an isolated voyage, fitted into a developing series of attempts to set up permanent Russian space stations – the Salyut space laboratories.

This project had started years before with a disaster. The astronauts who had lived in Salyut 1 in 1971 had died on their return to earth, because a valve jarred open on re-entry and let the atmosphere escape from the Soyuz spacecraft. The Salyut eventually burnt up. Salyut 2 went into orbit in 1973, just before Skylab's launch, and was no doubt intended to upstage it. But Kosmos 557, which was sent towards it, never docked, and eventually both it and Salyut 2 fell to earth and burned up. Kosmos 557, the Russians then said, was a trivial unmanned probe.

The Russians had to improve the Soyuz to make it more reliable, and this led to a couple of successful flights in 1973. Now they could resume their attempts to set up space laboratories. Salyut 3 was launched on 25 June 1974 and Salyut 4 on 26 December. Though the Russians have not mentioned the fact, not all Salyuts were identical: they were either research craft or military craft with two military personnel and a recoverable module that was jettisoned and recovered on earth – it presumably contained film. (One major use of a permanent space station is military surveillance.) Both craft stayed in orbit and were visited by astronauts in Soyuz vehicles.

The first successful crew in these flights – Pavel Popovich and Yuri Artyukhin – went up in a Soyuz on 3 July 1974 to Salyut 3. This was intended as a long-term residence and was a modified version of the previous Salyuts. There were solar panels pivoted to follow the sun, and there was an exterior TV that could be used to inspect the hull of the craft. The water on the spacecraft could be recycled and there was special gymnastic equipment – a form of treadmill and a rather splendid spring-loaded sweatshirt that exercised the muscles of the upper body while the astronaut was running on the treadmill. These devices would help resist the physical deterioration risked by a lengthy stay in space.

The astronauts worked through a programme of observations of the earth – geological and meteorological observations in particular. On 19 July they climbed back into the Soyuz and returned safely to earth. Meanwhile the Salyut continued in orbit, waiting for its next visitors. The first trip had been an outstanding success.

The next trip was less so. On 26 August Gennady Sere Sarafanov and Lev Dernin took off for a second rendezvous. They never made it. They landed back on earth at just before midnight on 28 August having completed what the Russian news agency described as experiment in orbital change. What seems to have happened is that the automatic stabilisation system failed and the astronauts used up their fuel in manual

147

Salyut orbit station in the workshop

control. It isn't very likely that they would have planned to land in darkness.

The Salyut was never re-occupied, but it continued to survey the earth automatically through TV cameras and sent the results back. On 24 January 1975 it was shut down by remote control and brought down to earth. It burned up, as intended, as it hit the atmosphere.

Salyut 4 was a more ambitious craft still. It carried a solar telescope and some other astrophysical instruments, but more interesting was the equipment which was clearly designed to explore the possibility of long missions. As in Skylab there was a bicycle exerciser, and also a rudimentary garden. Gubarov and Grechko went up on 5 January and spent thirty days in space.

The Russians, like the Americans, planned gradually to extend their space trips and, on 5 April 1975, Lazarev and Makarov set off for a two-month stay. Unfortunately the mechanism for jettisoning the second

148

stage of their rocket failed and was eventually separated from earth by remote control. By then the spacecraft could not hope to go into orbit and it started to fall, more or less out of control, towards the Altai Mountains. Fortunately the automatic landing systems all worked and the craft landed safely by parachute on a snowy slope. It started to roll down, but a parachute line snagged a rock. The astronauts were rescued, injured but alive.

On 24 May 1975 Pyotr Klimuk and Vitaliy Sevastyanov took up Salyut 4. This visit was a success and they stayed for sixty-three days, overlapping the Soyuz-Apollo link up. While the crew didn't achieve anything of great scientific interest, they had clearly mastered the problem of avoiding physical deterioration in space. Having made good use of the exercise bicycle in flight they walked briskly from their craft after touchdown. Salyut 4 continued in orbit uninhabited until February 1977 when it made a controlled fall safely into the Pacific Ocean. Salyut 5 had already been launched: the Russians appeared to be recovering their touch.

The flight of Salyut 5 did not constitute any significant step towards a permanent station, though it was certainly not a disaster. The craft was launched empty in June 1976 and Boris Volynov and Vitaliy Zholobov went aboard on 7 July. They worked on some biological experiments and they tried out weightless metal-processing, but failed to report any major discoveries. Clearly the aim was to stay longer than the previous Salyut record of two months: three months would be a logical next step.

After seven weeks the astronauts were brought home. The Russians gave no reasons for this at the time; they weren't even prepared to admit that the return was premature. The fact that the astronauts had to land in the dark, though, suggests that the decision to end the visit was hurried. A much later report mentioned that the astronauts were in a poor physical condition.

On 14 October Zadov and Rozhdestvensky were sent towards the Salyut, but the link-up failed. They returned hurriedly and landed, in the dark, in a near-frozen lake in Soviet central Asia. A third pair, Gorbatko and Glazkov, went aboard on 8 February 1977 and stayed for a modest eighteen days.

The turning-point in Russian attempts to establish a permanent laboratory in space came with Salyut 6, although it started badly. It was launched on 29 September 1977 and on 9 October Vladimir Kovalenok and Valery Ryumin were sent up to board it. They failed. The rendezvous was accurate and the Soyuz made several attempts to dock, but each time the latching mechanism failed to work. Alarmingly the problem was on the Salyut, not on the Soyuz which could have been taken back to earth for repair and replacement.

The profession of inflight spacecraft repairmen isn't crowded but the Russians had an ideal candidate, Georgy Grechko, a skilled engineer who had already been in space and who was in training for a later visit to Salyut 6. He was sent up with Yuri Romanenko on 11 December 1977 and docked with the Salyut. This made it clear that the Salyut had two docking ports: the first spacecraft sent to rendezvous must have run out of fuel before it could try this second port. The second port was essential for a station intended for long periods of residence. The idea was that the inhabitant astronauts should moor their Soyuz at one port, leaving the other free for the delivery of supplies or even for visitors. One of the ports in the Salyut was thought to be defective, so the astronauts docked at the other and started setting up the repair equipment. This took a week. On 20 December they were ready to look at the defective port.

Both astronauts put on spacesuits, then Grechko left the Salyut through an airlock and worked his way, hand over hand, to the suspect port. Romanenko stayed half in the airlock as a safety measure. Grechko investigated thoroughly then decided that there was nothing wrong with the docking port. Either the last Soyuz had been faulty or there had never been any fault at all, merely a failure in the instruments. What this meant was that the Salyut was available for long-term stays after all. Grechko was delighted to be able to report this to Mission Control. Then he prepared to get back into the Salyut. He was passing over the Pacific Ocean at the time. Very few people have ever made a space walk, and Grechko knew that he would probably never again have a chance of looking at the world from the outside of a spacecraft. So he stayed there, watching the world go by. Suddenly he noticed that Romanenko was also going by. The second astronaut had eased himself into the airlock to get a better view and had been caught out by the odd effects of trying to move when weightless. He had drifted gently out of the airlock and as he had forgotten to fasten his lifeline he continued to drift, moving slowly past Grechko. There was a serious risk that he would drift gently away into space, history's first astronaut to be lost. Fortunately Grechko caught hold of the floating lifeline and pulled his comrade back. They prepared to crawl back into the Salyut through the airlock.

Now there appeared to be another problem. According to the instruments the airlock was leaking. If it couldn't be fixed both astronauts were doomed: when the oxygen supplies gave out they would die. It would be possible, in theory, to work around the outside of the Salyut and get into the Soyuz they had arrived in, but nobody, there or back on earth, could think of a practical way of doing this. In fact nobody could think of anything valuable to do if there really was a leak, so they were told to assume that there was an instrument error and crawl back aboard. It turned out that that was all that was wrong. The astronauts could settle

The model of the Salyut–Soyuz–Progress scientific and research complex installed in the Cosmos pavilion in the grounds of the Exhibition of the USSR Economic Achievements

down to living in the Salyut, extending the period in space above the previous best of a couple of months.

A relatively small spacecraft – around 47 feet long and 13 feet in diameter at its broadest – the Salyut could carry only limited supplies. The plans included recycling some essentials, growing some food, and having the main bulk of provisions sent from earth. At this stage all that could be recycled was the water condensed in the air-conditioning system. Attempts to grow beans and peas failed.

The supply system from earth was marvellously effective. The first of a series of robot supply-ferries, Progress 1, left earth on 20 January 1978 and docked automatically with the Salyut. It brought, in particular, a new variety of food. The astronauts had tested their food supplies before they left and had accepted them. But, as the American astronauts had found, in space the food was insipid and unattractive. For nutritional reasons, and to boost their morale, they needed a change.

They also needed exercise. The Skylab trip had shown that, without exercise, astronauts not surprisingly became weaker and weaker: their muscles atrophied. The Soviet schedule called for a couple of hours of exercise a day, using the treadmill in conjunction with the spring-loaded sweatshirt. The astronauts disliked this. Their days were rather full with

151

experimenting and maintaining the craft, and two hours of rather boring exercise was an unattractive proposition. The Salyut was also rather unpleasant as a gymnasium. The air-conditioning was easily overloaded and the astronauts' clothes became soaked with sweat. They consequently neglected their exercise routines and were unable to walk when they landed on 16 March 1978, after a Soviet record of ninety-six days in space. They were giddy, they felt ill, and they had to be carried away on stretchers. Fortunately none of the damage was permanent and the Soviet doctors knew that a disciplined approach to exercise would avoid it in the future.

Salyut 6 was the world's first spacecraft designed for extended periods of occupancy. It had two docking ports. It had a chemical method for replenishing the oxygen in the cabin atmosphere – the moist air was passed over potassium superoxide – and a system for condensing moisture in the atmosphere and purifying it for re-use. The water, of course, came from the perspiration and exhaled breath of the astronauts. Salyut 6 was on the way to being an outstanding success. It was so secure and reliable that the Russians started a 'foreign guest' project. Astronauts from Iron Curtain countries – the first ever astronauts who were neither American nor Russian – were taken up for short stays in the Salyut. Quite rightly this was seen as a demonstration of the fact that the Russians had made living in space an almost routine operation.

The Russians would have been aware of the value of making that particular point clear – the 'guest' project was a Soviet first: at that time the Americans had reached only the point of agreeing in principle with the European Space Agency that Europeans could go up in the Space Shuttle, particularly in Spacelab.

The 'guest' project also provided a technical advantage for the Russians. While the Salyut would stay up for years and astronauts might spend six months or even a year in it, they couldn't simply leave their Soyuz tethered on the outside, waiting to take them home. The rocket fuels and the electric batteries would deteriorate and the craft would become unreliable. Visiting astronauts were taken up to meet the residents, and introduce some variety into their lives; they could then return in the moored spacecraft leaving their own behind. The first European spaceman was Vladimir Remek, a Czech, who was taken up in Soyuz 28 on 2 March 1978. He was followed by a Pole, an East German, a Bulgarian, a Hungarian, a Vietnamese and a Cuban. Though the visitors did carry out some experiments, their visits did not play any part in the serious space research. The nationalities, in fact, were worked through in Russian alphabetical order.

Meantime the Salyut 6 and its crew were still in space. Vladimir Kovalyonok and Sasha Ivanchenkov had left earth on 15 June 1978, intending to spend 140 days in space, a new world record. They made useful observations of the earth and the weather and they helped to perfect the techniques for restocking, to enable the robot spacecraft to dock automatically. The astronauts simply unloaded the stores and let the spacecraft travel, again as robots, back to earth. Kavolyonok and Ivanchenkov stayed for their twenty weeks without experiencing any great problems, and were in good physical condition when they landed – much better than Grechko and Romanenkov after only ninety-six days. Unlike their predecessors, they had been serious about their exercise schedule. In addition, as the Skylab flight had shown, a long stay in space was more supportable than one of intermediate length. The Salyut itself was still usable after their return, so now the Russians decided to go for a really long stay. Vladimir Lyakhov and Valeri Ryumin were sent off on 25 February 1979 for a space trip lasting 175 days – nearly six months. This was a stunning new record and suggested that there might be no limit to the time that could be spent in space. The two astronauts spent this trip alone as there was a failure in a planned visit by a Soyuz. On the other hand, the unmanned Soyuz which was to carry them back to earth docked perfectly. There was also a stream of robot supply-ships including a giant Progress 1 ferry that brought new supplies of rocket fuel; another brought a radio-telescope aerial that could be used for marvellously accurate observations. Unfortunately the aerial jammed in one of the ports, and Ryumin had to go round the outside of the Salyut to cut it free. One of the purposes of the trip was to investigate the possibility of repairing Salyut equipment in space; normally it was simply replaced. Repair proved perfectly feasible.

As there seemed to be no limit to the time a man could spend in space and the Salyut 6 was continuing to function well, the Russians sent up another crew, Ryumin (again) and Popov, on 9 April 1980. They stayed for 185 days, a new record time. In 1982 Anatoly Berezovoi and Valentin Lebedyev set another record when they spent more than six months in Salyut. Settling in space was clearly a real possibility.

The Russians are in a position to create the first space settlement simply by extending the Salyut programme. They are not disturbed by the expense of using 'throwaway' rockets and have disclosed no plans for a re-usable shuttle. They could launch individual units and join them up. They could build large stations, move them to a lunar orbit, and then send a near-robot craft with astronauts who could use the station as a base for a moon landing. In fact the ability to send supplies and people by automatic rockets opens up enormous possibilities to the Russians. They talk of going to Mars by the year 2000: their approach needs no further technological development, merely a painstaking duplication and extension of what they have already achieved.

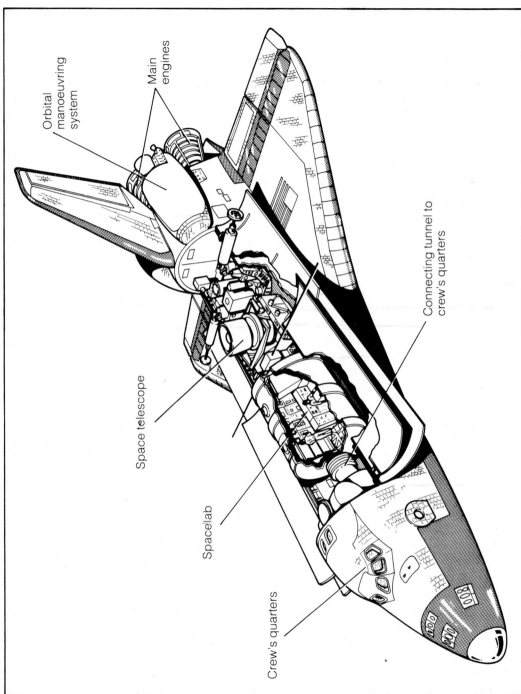

Orbital
manoeuvring
system

Main
engines

Connecting tunnel to
crew's quarters

Space telescope

Spacelab

Crew's quarters

Space shuttle

154

10

THE SHUTTLE ERA

In 1969 NASA seemed set for a marvellous future. It had put men on the moon ahead of the deadline set by President Kennedy, and the world had been excited by the achievement. It had developed considerable technological expertise, and it had the backing of the public. There were further exhilarating plans to be executed.

Then the financial roof caved in. America, at the time, was undergoing a period of political disturbance and urban rioting following the European political riots of 1968. Suddenly the public became disenchanted. While it was exciting to have been part of the nation that won the space race, there didn't seem to be any point in sending further missions, especially when it became obvious that the astronauts were not doing anything very much once they arrived on the moon. And with shortages of cash restricting so many useful projects on earth – projects that would directly improve the quality of life – the citizens of America began to feel that they could find better use for the money that went to NASA. The politicians reflected this feeling. NASA's proposed budget for the fiscal year 1971 was slashed.

Skylab was already well under way, but for future plans the space enthusiasts at NASA had to think of more economical ways of exploring space. It was partly a matter of keeping going until the economy improved again, but even so, once space research ceased to be seen as a titanic battle between the superpowers it would no longer command unlimited funds. NASA decided it would look at two projects: a space shuttle and a twelve-man space station.

The space station was to be built and launched after Skylab, which was a relatively temporary home in space. The new station would be manned, partly for military, partly for scientific reasons, on a more or less permanent basis. In other words, the plans had to include constant replacement of the observers.

The project was insanely expensive. Each rocket to Skylab cost $120 million and the launcher, once used, had no further purpose: it burned up in the atmosphere, it fell into the sea, or it joined the vast clutter of orbiting space debris. Skylab only had three crew changes and used only three launchers; the twelve-person station would need many, many more. The cost for launchers alone would be huge.

NASA argued that it would have to build a shuttle to service the orbiting station. The American Senate accepted the argument but decided that it didn't want either the station or the shuttle. There was a real risk that NASA would have no future projects to work on, but like other space enthusiasts, it had long learned that any project could become acceptable if its name and aims were framed in the right way. The shuttle was now proposed as a way of extending *unmanned* research and there was, NASA conveniently discovered, no immediate need for an orbiting space station. Unmanned research was relatively acceptable because it was so much cheaper, and there was even a real possibility of actually making a business out of it: the first communication satellite, Echo 1, put up in 1960, had hinted at the possibilities of what was to become a multi-million dollar industry. Finally NASA also managed to link the project with the needs of the US Air Force, which was fascinated by the possibility of a plane that could travel in space but land on the ground.

One purpose of the shuttle is to put satellites into orbit. Space satellites have proved their value: the earth has been mapped more accurately than ever before using pictures from the Landsat series of satellites. These circle the earth in an orbit that goes close to the poles so that the earth spins beneath them; they then survey the earth with cameras that are sensitive to light of particular colours and to infra-red light. They transmit the results of the surveys to earth, where they are converted to pictures of great beauty as well as geographical accuracy. We also have satellites that survey the earth for the information that the weather forecasters need, satellites that look for geological evidence of valuable minerals and other natural resources, and satellites that survey the crops of the world, to give warning of gluts and famines. There are also communication satellites. At present these are the only genuinely commercial satellites: they are owned by business corporations that make money by renting out the facilities of the satellite. The benefits they bring include transatlantic telephone calls and international radio and TV programmes: information is easily moved from one side of the globe to the other.

Satellites are also used for military purposes. Every country is in a position to acquire information about every other country's state of military preparedness, or lack of it, from satellite observations. There are satellites circling near the earth, taking pictures so detailed that the registration numbers of cars are visible; they look for changes indicating that a new military base is being developed and, during actual fighting,

FACING PAGE
above, *Mission control during the Voyager 1 mission*

below, *Emergence of a satellite from the payload area of the shuttle (STS-5)*

156

they can follow the progress of battles, rather as people used to do from tethered balloons in the First World War.

FACING PAGE
*The first launch of US
space shuttle Columbia*

The military certainly intends to use the space shuttle which was eventually built and it has booked space on at least the first fifty trips. One of its aims will be observation, and as the shuttle can put a satellite into orbit for $18 million – a shuttle flight can take up a number of satellites – compared with $100 million using the 'conventional' Titan rocket, it will be observation at bargain rates. In any major conflict both sides will rely heavily on observation satellites, so a second function of the shuttle will be to put enemy satellites out of action. The shuttle can be fitted with an enormous, 50-foot manipulator arm, which is designed so that it could move among the satellites, putting those of the enemy out of orbit. However, the method is likely to be somewhat more subtle. A spy satellite must keep very precisely in its orbit if it is to send back an accurate picture, and it does this by locking on to a 'space mark' – perhaps a star. It is fairly easy, with a powerful blast of radio waves or laser beams, to destroy the sensors that lock on to a star and the super-powers are undoubtedly studying ways of sending up 'killer' satellites that home in to the enemy's spy satellites and destroy their equipment. At least one Soviet satellite and one from the USA have crashed mysteriously.

Although the space enthusiasts have had to nurture military enthusi-asm in order to raise money for their projects, the shuttle will also serve the civilian community, particularly by retrieving and repairing ailing satellites. It is costly to put a satellite into orbit even using the shuttle, so a satellite that fails represents a lot of wasted money. The Orbiting Astronomical Observatories (OAOs), for example, were a very expensive failure. Four of these were launched during the 1960s and early 1970s. One failed because an electrical battery didn't work; another because one of its two observational instruments broke down; the shroud that held a third to its launching rocket didn't come clear, as it should have done. All these problems could have been prevented or solved if the OAOs had been placed in orbit from a shuttle travelling in a very similar one: NASA estimates that this would have saved between two and three hundred million dollars. The total cost of a satellite flight, at today's prices, is between thirty and thirty-five million dollars. Just one repair job on the OAOs would have justified a great deal of research and development.

The idea of a space shuttle predated NASA's interest. During the Second World War the Germans, Dr Eugen Sanger and his wife-to-be, Dr Irene Bredt, designed an enormous long-range bomber. The fuselage was slim, square in section and 92 feet long, although the wings were stubby and carried smoothly across the underside of the fuselage. The aim was to

157

bomb America from Europe: up to that time there had been no possibility of making a bomber that could carry a functional load that far. Once the plane had been hurled upwards into space, it would not need any power to keep it going, so it was placed on the trolley of a rocket-propelled monorail that shot up a slope and stopped, leaving the plane to travel on into space where it would be powered by its own rocket engine. The monorail trolley dropped back down the slope to be used again. The plane was to travel above the atmosphere in a series of skips, like a stone skimmed across water, until it was near enough to the target to come down through the atmosphere and release its bombs. Technically the craft would be re-usable, although in practice it would make only one-way trips into enemy territory. In fact there was no practice, because the project was abandoned in 1942.

This was the year of the first launch of Wernher von Braun's A-4 rocket, eventually to become the V-2, which was used to bomb Britain. With only slight modifications this rocket later became the Redstone, one of the first US space rockets. As an offensive rocket its range was limited, so von Braun had the idea of fitting wings to produce a skipping, long-range bomber. An expert in the art of getting backing for his space projects, he called his skipping glider the A-4b so that he could use A-4 funds in its development. However, the plane never flew in earnest.

Neither of these craft was used, because their inventors had failed to solve the problem of steering them. They could be directed like a rocket and, once they were well down in the atmosphere, travelling as an aircraft rather than a rocket, they could be manoeuvred like an aircraft. The same did not apply in the upper atmosphere, where the air was so thin that ailerons, elevators – movable control surfaces on wings and tail – and rudders had nothing to press against; and it certainly wouldn't work in space. The solution to this problem came from a series of US rocket planes that was at one time proposed as the basis for the MISS programme to put a man into space. The series was initiated with the X-1, which, on 14 October 1947, became the first plane ever to fly faster than sound. It was followed by a series of X-planes, designed to fly at increasingly high speeds – the X-15 could fly above 4000 mph – all using conventional aircraft controls in the denser parts of the atmosphere and small control jets for manoeuvring in the thinner parts. The X-series planes were launched from high-flying planes – they couldn't carry enough fuel to climb from the ground. A later series, codenamed ASSET and PRIME, used Atlas rockets to take them even higher, where they could fly even faster. Flying at around 12,000 mph, these planes could investigate the problems of control during re-entry from space and of flight through the upper atmosphere – the shuttle would later need solutions to both of these problems.

The Atlas booster that took these aircraft aloft was not, of course, re-usable. It seemed at the time that a re-usable booster would have to have wings so that it could be flown back to earth, either by a pilot or by remote control. This realisation led to some odd designs. The British MUSTARD project, for example, used three winged rockets, each with a pilot. All three fired at once to get the assembly off the ground, but only one went into orbit while the other two were flown back to base. It seemed a good idea, as did Boeing's Dyna Soar, which was to be taken into space by a Titan rocket, to orbit the earth on a reconnaissance mission and then glide back. Neither of these projects was actually developed, but they both helped to define the eventual design of a shuttle. So did the studies of 'lifting bodies' – aircraft built without wings, but shaped in such a way that their speed through the atmosphere generated lift in much the same way as a wing does. The problem, always, was to design a craft that would survive re-entry – which would tend to tear off any wings – and yet be controllable in the lower atmosphere.

In the early days before the budget cuts NASA had asked various aerospace companies to design a re-usable shuttle: it wanted a two-stage, wholly re-usable craft. There was to be a big piloted booster that would take the manned orbiter aloft to manoeuvre in space with its own rocket engines. A flurry of designs was proposed but they all stretched technology, and credibility, to its limits. The orbiter had to be very light and the booster very powerful, and all the designs were marginal: if the actual machines fell at all short of the design aims, the shuttle wouldn't work. In 1971 NASA eventually abandoned the idea of using a manned booster. The following year, with President Nixon's blessing, NASA started on the design that was to lead to the space shuttle. After a quite extraordinary series of delays and problems this finally made its first successful flight in 1981.

The projected shuttle was to consist of an orbiting spacecraft called, logically enough, the orbiter. This had small fuel tanks for manoeuvre and landing, and a gigantic external fuel tank that could be discarded once it was empty. As the orbiter's engines didn't have to propel a large empty container they could be relatively low-powered. The orbiter and tank were to be boosted towards orbit by a couple of re-usable engines and solid-fuel rockets were eventually chosen. It was a fairly straightforward task to get the orbiter and fuel tank off the ground, so they could be much simpler and hence cheaper than liquid-fuelled motors, which need complex systems of valves and control. If the solid-fuelled rockets did, by accident, fall into the sea and sink, the loss wouldn't be disastrously expensive. The orbiter itself would not have motors: it would be a glider controlled by thruster jets. This meant that the pilot had to get

160

the tricky, fast, steep landing right: there would be no way of circling round for another attempt.

Gradually the design of the shuttle crystallised, much of it taking account of the need for economy and re-usability, until it finally reached the form that was launched in 1981. Apart from the discardable boosters, the shuttle has three main rocket engines clustered in its tail. These burn hydrogen and oxygen, which are carried in the large external tank and, with the boosters, carry the orbiter almost into orbit. They are a triumph of engineering because, like the rest of the orbiter, they are designed to function for fifty or even a hundred missions, a total of several hours of burning. The hydrogen and oxygen fuel is used once only, to leave the earth. For manoeuvring, including the first manoeuvre of going into orbit, the orbiter carries its own fuel, a hypergolic mixture – one that ignites when mixed. Space engineers use these exotic fuels where a failure to light immediately would be disastrous. In the shuttle they are burned in two of the main manoeuvring engines near the tail, and in arrays of small thruster engines, at nose and tail, for delicate movements in space. The orbiter itself looks like a very swept-wing aircraft; it measures just over 122 feet long, with a wingspan of 78 feet. The astronauts who fly it, and the crew who work on it, all travel in the forward section.

In many ways the shuttle launch marks the dawn of the true Space Age – an age when space will become a familiar domain – and one way was in the 'democratisation' of its crew. While the plane will be flown by astronauts of the traditional type – highly trained, rigorously selected people – the crew that operates the equipment faces much lighter demands. And when the shuttle takes scientists into space to experiment, these payload specialists will face an even less rigorous selection. Any fit person – man or woman – will be able to go.

In the forward section the astronauts and crew live and work in normal clothes, breathing air at sea-level pressure. To the rear of this is the gigantic payload bay, 60 feet long and 17 feet wide, with doors that have to open in orbit so that the orbiter can get rid of excess heat. The payload bay can carry satellites to be placed in orbit, satellites that have been recovered from orbit for repair, scientific instruments for space experiments or, on some of the flights or Spacelab, a sealed laboratory containing space scientists and their experiments.

The space shuttle takes off like a rocket and lands like a plane. For its launch it sits on its tail, with the two solid-fuelled boosters on either side, all dwarfed by the gigantic external fuel tank. The boosters and the orbiter's engines are used for take-off. Once the solid-fuelled boosters are burnt out (it takes two minutes), they drop away and parachute to earth for recovery. Soon afterwards the external fuel tank, now empty,

opposite, *The launch of the first US space shuttle, Columbia* is discarded and falls to earth, though as yet there are no plans to recover and re-use this. Shortly after this the orbiter uses its manoeuvring engines to put itself into orbit.

When it is time for the orbiter to return to earth, it has to slow down so that it drops out of orbit. Its engines can only drive it forwards, so to slow the shuttle must turn so that it is flying backwards and then fire its engines. It finally turns again so that it hits the atmosphere flying forwards, to land as a very fast glider at about 200 mph.

Striking the atmosphere from space generates an enormous amount of heat, and earlier spacecraft used a heatshield (known as an ablative shield) that burned away. The orbiter is protected from the heat of re-entry by a layer of insulating tiles. The problems of designing tiles that would withstand re-entry temperatures of up to 1200°C and of fixing them to the complex curves of the orbiter bedevilled the project: some of them even fell off during the first flight, although fortunately without any danger to the orbiter or its occupants.

The original purposes proposed for the shuttle mainly involved satellites, but it will also take up devices for research. One of the most promising is the space telescope, a gigantic, 94-inch optical telescope which will be able to see the stars by the light they emit, in the same way (though on an infinitely more powerful scale) as the much larger – 200-inch – telescope at Mount Palomar in the USA. The space telescope will be taken up in the shuttle and placed in orbit: it will be controlled from earth and will send its pictures back there. One function of an astronomical telescope is to focus the light collected from a faint star. As more light falls on a large telescope than a small one it can see fainter stars; and, as it is a reasonable assumption that the fainter stars are the more distant ones, this means that it can see further into space.

On earth, astronomers are hampered by the fact that the light that arrives here is scattered by the atmosphere – otherwise our sky would be black, as is the sky, day and night, above anyone in space. This scattering means that some light simply never reaches any telescope here, and the light that does, even though it comes from a tiny point in space, is diffuse. Thus there is no way of seeing a very distant object accurately: even the best of our telescopes can see only stars that are 2000 million light years away, i.e. light that set off 2000 million years ago.

The space telescope will be able to see 14,000 million light years into space, because it can be very accurately directed by locking in on to a star, and because it doesn't have to look through an atmosphere. It will of course be possible to see stars and other objects in the sky that we have never seen, but there is also a much more fascinating prospect. We believe that the universe was formed about 14,000 million years ago. If the space telescope is pointed in the right direction it will act as a

opposite, Underview of orbiter Columbia showing the thermal tiles which are essential for re-entry into the earth's atmosphere

miraculous time machine: we might be able to see the light that set off at the time the universe was formed.

Nearer home, there are equally fascinating observations to be made. Most people believe that there are some forms of life in other places in the universe. It seems that life on earth came about because the earth is hospitable – various chemicals that are fairly easily formed from substances that exist throughout the universe were able to develop into a very simple form of life on earth and from that moment evolution took over.

The same thing could happen elsewhere in space, and simply because there are so many stars, many of them with planets, there must be some planets that are as hospitable as earth, though they might vary sufficiently to encourage different forms of life from our own. It would be fascinating to investigate those planets, but at present we know of no way of locating them. We are now pretty sure that there is no developed life on any of the planets of our own sun, but it would be interesting to know about planets going round other stars. (In reality, a planet doesn't travel in an ellipse around its star: both travel around a common point.)

Even in the best telescopes, these stars appear only as points, so we cannot hope to see a planet going round, even though it could be much the same size as the earth (8000 miles in diameter), and the star will be sun-sized (more than a hundred times as large as the earth). However, as a planet goes round its star it makes the star quiver. The marvellous accuracy of the space telescope can detect this quivering, enabling us to locate remote planets. Once we know where some of these are, we can look for traces of life there, or even try to pick up radio signals from their inhabitants.

After the first few trips the space in the shuttle will be available for rent by anyone with experiments to conduct. A major group of experiments will be conducted on the shuttle's tenth trip, scheduled only tentatively for late 1983, using an orbiting laboratory called Spacelab for the first time. Though Spacelab will be used for scientific research, its fascination is as a testbed for the exploitation of space. It is in Spacelab that people will try out proposals for space industries.

Though a version of Spacelab was proposed in 1969, the current version is based very closely on the experience of Skylab. Like Skylab it is an orbiting laboratory, but at present it is planned that Spacelab should go up and down on one shuttle and not stay floating in space. Eventually it might be left in space, as Skylab was, but for the present its experiments will have to be fitted into the week or so that is the orbiter's limit.

Spacelab is a European venture, built by the European Space Agency (ESA); it cost about £500 million, which is relatively cheap for an entry into space research – gloomy Britons have noticed that this sum approximates to the average annual loss of some of the less successful nationalised industries. The money for Spacelab is provided by a group of European nations on a very uneven basis. Germany put up the most money, Britain only a small fraction.

The laboratory is built in two detachable parts – a sealed can 4 metres across and a mere 2·7 metres long, which will give the researchers a shirt-sleeve atmosphere, and an open pallet that can carry experimental equipment to be used in space. The payload bay can carry two joined cans (to give the payload specialists some elbow room) with one pallet, or one can and a longer pallet. The whole space could also be taken up with experimental pallets, with the instruments operated and controlled from the forward section of the orbiter.

Spacelab is attached to the payload bay of the orbiter and connected to it for its supplies. But it is a workshop, not a living space, and the scientists who work there will float in from the orbiter's crew area along a tunnel. The NASA people who designed Skylab had the novel idea of using some working space to the utmost by having benches radiating from the centre line, rather like the spokes of an upright wheel. The researchers, they argued, could float from bench to bench. This didn't prove practicable. The astronauts in Skylab, well trained though they were, had spent most of their lifetime in normal gravity where benches were horizontal, and they were disoriented and confused, even made slightly sick by the arrangement in Skylab. Spacelab's designers recognised that their laboratory would have to be suitable for people, rather than fitting the abstract logic of an engineer.

The plans are to take Spacelab up on the shuttle's tenth and fourteenth flights, but it will eventually be used more regularly. About eighty of the shuttle's first two hundred flights will take Spacelab, assuming that anyone wants to use it. It can be rented for research for about £7 million per flight but the user will also have to pay the £18 million that a shuttle launch costs.

The first two trips will be tests, to make sure that the system works. There will be a total crew of six – two astronauts, two mission specialists and two payload specialists – to give two teams of three working in twelve-hour shifts. It is fundamental to the working of Spacelab, as to the shuttle, that specialist scientists go up as payload experts: a scientist in space can monitor the experiments, or alter them, or even redesign them if there is some serious problem.

Spacelab's experiments will include investigating potential space industries, for which plans have not progressed very far since the days of Skylab. Industry has shown a stunning reluctance to suggest new space projects, let alone put money into research, but this is probably a

165

temporary phase. The whole idea of living and working in space is new, and people's imaginations are blocked by the novelty. Once the possibility has become a reality the ideas will follow.

Not all of Spacelab's experiments will be on a grand scale: NASA is encouraging people to devise small experiments that can be packed into odd corners of the shuttle. These will be taken up for the bargain price of £20.00 per pound, very little more than air-mail charges. However, the experiments will have to be very carefully designed to justify a ticket. If they involve life – experiments with beans and spiders have been proposed – the researchers, many of whom will be students, will have to devise a way of providing an atmosphere. And all the experiments must be automatic and self-recording as the inventors aren't going up with them. Experiments to see how liquid alloys solidify, how beans grow and throw out roots, and how spiders spin their webs in the weightlessness of space, are among early proposals.

The shuttle's trips cannot last more than a week or so because it doesn't carry enough energy to stay longer in orbit. It therefore doesn't have to provide the comfort of permanent space settlers, although it is nevertheless much more roomy than, say, the capsules that went to the moon, and even more so than the earlier two-person Gemini spacecraft.

The astronauts and the experimenters live and eat and cook in a low-ceilinged room that is in the lower part of the forward fuselage, under the flight deck. The latter is a marvel of technological complexity – there are more than 2020 separate displays and controls, although this figure is somewhat misleading in that there are separate sets of instruments and controls for the pilot and co-pilot, who sit side-by-side. There are also seats for two other crew members. In the early days at least, one of these will be what NASA calls a mission specialist – essentially someone who works the devices on the shuttle while the astronauts get on with their task of piloting. The fourth seat will be for a payload specialist, normally a scientist or an engineer in charge of a particular experiment, although the rich adventurer who is reputed to have paid for his own flight on the shuttle, and the science-fiction writer who is reported as having had *his* passage paid for by his fans, will each have the opportunity to ride in the payload specialist seat.

Off duty, the astronauts and crew will sleep on bunks in the forward section. At least one of these is vertical, or would be described as such if the orbiter were flying horizontally above the ground: the other three are, in the same sense, horizontal, although the base of the bottom bunk is 'above' its occupant; if the plane weren't in space the occupant would fall out.

The people on board the shuttle work to Houston time: they get up at around six, dress in their light blue uniforms, and then put on the suction shoes which anchor them to the floor or, come to that, to the walls or ceiling. They float to the bathroom, using the handholds around the cabin: without these, the crew might simply ricochet slowly around like spongy billiard balls. Both the lavatories and the washbasins have suction pumps to draw sewage into storage tanks that will be emptied back on earth. Unlike the sewage systems of some earlier spacecraft, those of the orbiter won't pollute space.

Clean and ready to eat, the crew will drift over to the galley. The food available for breakfast, and indeed for all meals, is plentiful and varied, although NASA has had to discontinue the practice of letting astronauts choose their own menus: previous spacecraft returned with a mass of the least popular dishes. Most of the food is dehydrated. There is no shortage of water in the orbiter because its electricity comes mainly from fuel cells which produce energy by allowing hydrogen and oxygen to combine: the by-product is water. A typical breakfast will consist of branflakes, scrambled eggs and beef pattie, together with a preserved peach and some cocoa or orange juice. A few foods are taken in the normal form,

Flight-deck of Columbia

but these are mainly bread and such traditional expedition stand-bys as beef jerkie – sun-dried beef which is popular, it is alleged, among cowboys. Getting a meal ready is simple. One person prepares the food for the whole crew. The dehydrated foods are mixed with water, and those that are to be heated are put into an oven; the whole meal will then be spread out on airline-type trays. It will be nutritious and varied, though there is no risk of the orbiter getting any recommendations in the Michelin guide.

The crew's working days will be quite busy – and they will tend to become busier in the later flights as NASA's confidence in the crew's abilities grows. The orbiter is not only the first re-usable spacecraft: it is also the first in which the crew have control. In earlier types of spacecraft the computers and many of the instruments were on earth, at Mission Control, and very often the astronauts had no freedom of choice – they did what NASA instructed them to do. The orbiter has its own full set of computers. The astronauts are responsible for checking the fuel supply, the state of the craft's atmosphere, and other parts of the life-support system. They can also plan the actual patterns of the flight, or at least programme the computers on board to do it for them. As the orbiter will spend most of its flight in orbit, and as there will be plenty of devices to give warnings if anything goes wrong, the astronauts will have time to help the mission specialists and payload specialists with their tasks.

On the first few flights the main tasks will involve launching satellites. One flight has already been booked for three communications satellites. They have to be placed in an orbit much higher than the orbiter's so that they travel round the earth at the same speed as it rotates and will therefore seem to be stationary. This is called a geosynchronous orbit. Each satellite will be fitted with rockets that will take it into orbit and keep it in position there, although there is a plan to build a space tug that will haul batches of satellites from orbiter heights to the geosynchronous one.

A later flight will take up a number of experiments on a platform, deposit them in orbit and then, when the experiments are complete, retrieve them: this will be a space 'first'. The first thirty to forty trips are already fully booked, although the shuttle will obviously be available for unpredictable emergency tasks. Eventually it will be used to build space ports and power-satellites and to ferry people and materials into space. It may help to build settlements and then to supply them.

The space age – the true space age of the shuttle – should have started on 10 April 1981. After an enormous number of delays – the maiden flight was years overdue – Columbia, the first orbiter, stood on its launch pad at Cape Canaveral, flanked by its two solid-fuel rockets and overshadowed by the enormous liquid-fuel tank. The two astronauts, John Young and Robert Crippen, had been aboard for some hours, strapped horizontally, wearing their spacesuits and watching the instruments and computer print-outs that recorded the progress of preparation during the countdown. The press, radio and television of the world was ready; so were the droves of spectators from the companies that had helped to make the shuttle. The countdown reached nine minutes before lift-off, then there was a planned delay. It was followed by an unplanned delay caused by computer troubles.

Space travel is a product of the computer age – there is no human way of making the calculations involved in space travel minute by minute. Columbia was fitted with four on-board computers that would monitor the state of the spacecraft, tell the crew what had to be done and, when required, actually do it. There were only two operations that the crew had to perform without computers – lowering the undercarriage to land, and braking once the craft had landed.

The four computers were not isolated units. They all took part in all decisions so that if one failed or miscalculated, the others could carry on. They 'discussed' the decisions with one another and sometimes even disagreed, in which case the programming told the four always to act on a majority decision. A fifth computer was added as a referee.

On Friday 12 April this fifth computer failed: it disagreed with all the others. In fact, it didn't seem to be getting any messages at all from the others. Columbia couldn't start a journey with one dud computer, so the experts tried to find the fault. They re-ran programs, re-checked readings, and could find nothing wrong.

Time was passing. The astronauts lay strapped down. The safety rule was that they shouldn't spend more than six hours horizontal before lift-off. And they had been awake since early morning: they were not allowed to have a longer first day than twenty hours. Computer problems that involve either the machine or the program – no one knew which at the time – are immensely complex and this one would, it soon became clear, need time and people. Eventually around a hundred computer experts were summoned to work on the problem. The launch was postponed.

Shuttle launches can't be postponed by a day at a time. The liquid-fuel tank has to be emptied and, as this is maintained at the temperature of liquid hydrogen, it has to be left to settle to a new temperature before it can be refilled. The earliest possible date for another try was two days later, Sunday the 14th. Even that depended on solving the computer problem.

The solution was actually a surprising one. Computers function by directing tiny pulses of electricity around their circuits, and these are directed and synchronised by what are essentially clocks. Obviously, if the clocks get out of synchronisation on connected computers, communication fails. The clock on the fifth computer was doing just that. When the computers were switched on, the fifth one might start in time, or it might start slightly out of time. To solve the problem the computer engineers switched it on and off until it was running properly, and then left it running until lift-off.

The second attempt was scheduled for the early morning of the Sunday. It was a fine day and the crowds, even larger than Friday's, waited nervously through the countdown, applauding as the critical computer check passed. There was no other problem. At 7 a.m. Columbia made a dramatic and memorable take-off, faster than the earlier Apollo craft, leaving behind a thick cloudy stream of white exhaust from the liquid-fuelled motors mixed with the black smoke from the solid-fuel boosters.

Exactly to schedule, two minutes into the flight, the solid-fuel rockets dropped away. They fell into the Atlantic Ocean and were duly picked up by a couple of specially designed recovery ships. As is customary, there was also a Soviet trawler in the area: it was warned off from picking up the rockets by the US Navy.

This was the first time solid-fuel rockets had been used in a manned space trip, and it needed the biggest rockets of this kind ever built. It was also the first ever recovery of any space rocket with a view to re-use. Whether these rockets are actually used again will depend on how well they have stood up to their flight and their immersion in sea water. If they are too severely damaged the quality of future rockets will be improved.

Six minutes later in the flight, the enormous liquid-fuel tank was discarded, as intended, to fall to earth. It was given a toppling motion as it went so that when it struck the earth's atmosphere it would continue to fall, not skip along the top of the atmosphere and fall unpredictably. It broke into pieces as it fell and dropped into the Indian Ocean. It is the only major part of the shuttle system that is not designed for re-use, although there are plans to use fuel tanks, or modified versions of them, in space structures.

The pilot of the Columbia, John Young, fired his orbital manoeuvring motors briefly at this moment, to move the orbiter clear of the fuel tank. They then were used again to put Columbia into orbit. The first part of the mission was complete. The craft, free of the motors and the fuel tank it had needed to lift off against the earth's gravity, was lapping the earth briskly.

The take-off hadn't been perfect. Once Columbia was in orbit the crew used a television camera on the tail-fin to survey their craft, and they quickly saw that some of the tiles that had been cemented on to

170

shield the spacecraft from the heat of re-entry had been shaken off by the shock of take-off. More than a dozen were missing from the engine pods at the rear and there also seemed to be some missing from the nose. The loss created no great concern. The craft would eventually strike the atmosphere the right way up so the upper surfaces, those the TV camera could see, would not be so strongly heated as those on the underside.

This raised the next question – were any tiles missing from the lower, more vital area? There was no way that Young and Crippen could check this from inside Columbia. They could, of course, leave the craft and float below for an inspection, but there was another strategy to try first. The US Air Force had, for undisclosed defence purposes, set up a very high-powered telescope system that could examine Columbia as it travelled past. This showed that there were no tiles missing from the underside. Obviously some might fall off during re-entry, but this would cause problems only if some vital control mechanism were exposed to overheating: because the systems in Columbia were always backed by reserves the risk wasn't serious. There was no real likelihood of so many tiles falling off that the astronauts themselves would be exposed to a dangerous amount of heat.

This was a test flight and the astronauts occupied their time by checking the machinery, particularly the control rockets and the doors of the payload bay. These latter were vital. During flight they had to be open so that the craft could radiate excess heat, but they had to close securely for re-entry. There was always a risk that they might be warped by the intense heating they would experience in space and the planners had worked out various manoeuvres to deal with this contingency. One of these was the barbecue mode, familiar from the Apollo 13 trip: the orbiter was to be rotated so that the craft would be evenly exposed to the heat of the sun.

In the event, there were no problems. When the time came, John Young turned Columbia, 170-odd miles above the Indian Ocean, so that it was travelling backwards. He fired the engines so that the craft slowed and started to drop out of orbit, then he turned it again so that it was flying forwards, nose up: the underside of the plane was to take the shock of re-entry, 400,000 feet up, above Guam. For a short period the ionisation caused by the intense heat of re-entry cut all radio communication with Columbia. When it resumed it was clear that the craft was returning to earth exactly as planned.

The orbiter is essentially a mildly powered glider, landing steeply and fast. To lose speed John Young flew through a pre-planned S-turn. Later he was to say that this part of the descent had been controlled manually. The automatic pilot was fine, he said, for the straight parts, but there was a chance that it would be a bit brusque on the curves.

The fourth space shuttle – STS-4 – returning to Edwards Air Force Base after completing its mission in space

At 10·20·52, Pacific Time, on Tuesday 16 April, Columbia made a perfectly smooth landing at 200 mph, exactly where intended, at Edwards Air Base in the Mojave Desert of California. For the first time in history a spacecraft had gone up like a rocket and come down like a plane. For the first time a re-usable spacecraft had been tried. And, with only minor problems once the craft had been launched, the whole complex technology involved in cheap space flights had worked perfectly. If the term space age implies that space is part of man's territory for exploration, this trip, and not the moon landing – vastly expensive and relatively unsophisticated – signified its start. As Robert Crippen said, 'What a way to come to California.'

Though the first trip was a success, later ones have had problems, making it necessary to change plans. For example, a defect in the spacesuits meant that a spacewalk had to be cancelled. On the fourth trip, the shuttle achieved one of its commercial aims: it put communication satellites into orbit and thus earned some money for the first time. Even though it has had a troubled start, it is clear that the shuttle will eventually achieve its targets.

11

THE FUTURE OF SPACE

Space enthusiasts have always planned that man should eventually be able to live in space. Where it has been possible to convince the holders of the public purse that there is some military or – in the case of Russia – propaganda value in the establishing of at least small settlements, the experimental beginnings have been backed. The Russian approach – disposable rockets thrown away after launching – was expensive, but the expense was accepted.

In the West, where public spending is influenced by public opinion, it is the space shuttle that has unlocked the door to living in space. Without the shuttle trips into space were frighteningly expensive for the same reason that air travel would be expensive if you threw the aircraft away after a single trip. While the cost of the Apollo flights was borne as a matter of national pride and ambition during the moon race, afterwards it quickly became so unacceptable that the series was curtailed. Skylab continued because the military were interested, and the shuttle was acceptable because it could perform useful tasks, such as servicing satellites economically, and because it also had military possibilities.

As the space shuttle developed, it became clear that space settlements could be in existence by the end of the 1980s: only slight technological developments were still necessary. By AD 2000 there could be large space islands and miners working in the asteroids.

The cost of the Apollo series had antagonised the public because the people who were going into space were an élite, highly selected and expensively trained. The rest of the populace, who were financing the expeditions, began to feel that they were paying for adventures for a small group of superhumans. The fact that the astronauts demonstrated their humanity by playing golf on the moon, or smuggling food into the capsules, didn't help – that was no justification for manned space research.

Shuttle trips, on the other hand, will be able to accommodate anyone who is reasonably fit, which makes them at bit more acceptable to the

174

taxpayer, and eventually these ordinary people will be eligible for settling in space. They won't go there just for adventure; the people who will go into space to settle there will go to do a job. They will migrate to space as people once migrated to India, Australia and California.

In fact, space migration paradoxically offers the only hope for a future on earth. When man first went into space, people suddenly began to accept that the earth was finite: none of its resources would last for ever. For years people had reassured themselves by noting that every warning of shortage had been a false alarm. Now they admitted that the amount of raw materials available must be limited. No doubt agricultural productivity can be improved and, if we are prepared to eat some of the more unusual proteins that industry proposes, the present population of the world could be fed adequately for a number of years to come. But the population is increasing dramatically. Unless the present methods of food production are galvanised, there is no chance of there being enough food, let alone enough buildings, transport and tools for everyone in the year 2000.

The crucial shortage will be energy. If we have enough energy we can manufacture fertilisers and pesticides, we can cultivate difficult land, and we can extract metal from ores which would otherwise be worthless. But energy is becoming increasingly expensive. We are having to look very hard for new reserves of oil – in the Arctic and under the oceans, for example – and we are also investigating 'renewable' sources of energy: windmills, geo-thermal heat, solar energy and hydro-electric power. These sources of energy will all become cheaper and more readily accessible as technological advances are made. The total annual energy consumption of the United States equals the total energy produced every year by the sun shining on a surprisingly low number of acres, but because the sun shines only in the daytime and is low in the sky for quite a large part of that time, it is expensive to collect solar energy in a form we can use. Transporting the energy to wherever it is needed increases the cost enormously. You could no doubt set up a solar energy plant fairly easily in the desert around Las Vegas, but transmitting the solar electricity into Los Angeles, Paris or London would require a lot of expensive transmission cables.

Other forms of renewable energy have the same problems. Wind power is less reliable than solar energy; tide energy is obviously localised; and the same applies to wave energy which can be profitably harnessed only where there are consistently high winds and a wide expanse of ocean so that the waves can build up to great height. These alternative sources of energy will all help to alleviate the shortage, but they won't provide energy so cheaply that it can be used to increase our supplies of fertilisers or raw materials.

But it will be possible to harvest energy from space where the sun is always shining and there are no clouds or rainfall: a giant solar collector could operate as an artificial satellite. The solar cells – the devices that convert the sun's energy into electricity – would be as expensive as if they were intended for use on earth, but they would collect and convert far more energy in space. And, because they would be weightless, the panels of cells could be supported on a very flimsy structure. The electricity would still have to be transmitted back to earth, but this would be simpler than sending it from place to place on earth because it would be beamed down as microwaves rather than carried by cable. The solar satellite would have to be stationed over the equator, 23,400 miles up, so that it appeared to hover in the same place in the sky. In reality it would be going round the earth at the same speed as the earth rotates, i.e. in a geo-stationary orbit. Such orbits are already used by communications satellites.

A solar-power satellite was first proposed by Peter Glaser, a scientist at the Arthur D. Little Research Institute. With some backing from NASA and help from the Grumman Aerospace Company he worked out a possible design. His satellite was to have two large rectangular panels, rather like wings, each three miles long. At the centre of this vast structure there would be a mile-long section for the equipment that converts the energy to microwaves, and an aerial 3000 feet in diameter to direct the waves to the earth. On earth the whole satellite would weigh 25 million pounds – more than ten thousand tons – but it would in fact be built in space where its structure could be delicate and elegant.

Any kind of radio transmission involves sending energy from one place to another by radio waves, but the amount of energy arriving at any one aerial is tiny. We don't know very much about transmitting large amounts of energy by microwave. In California there has been a series of experiments where enough energy to operate electric light-bulbs was sent half a mile by microwave. The scientists involved with these experiments don't anticipate encountering any major problems when they increase the scale – even from half a mile to 23,400 miles!

One problem is to keep the beam concentrated in order to maintain its strength: if the microwaves fan out, they won't supply a usable amount of energy. More importantly, the heating effect of microwaves makes them dangerous to life. Microwave ovens, which generate a lot of heat,

---

Boeing conception of a thermal engine power satellite being constructed in low earth orbit. The domelike structure attached to the main body is a partially completed thermal cavity – a solar furnace into which the sun's rays eventually would be focused. The large flat structures around the cavity are radiators through which gases used by the cavity's generators would be cooled. Solar reflectors would then be attached to the surrounding skeleton.

can damage people if the rays scatter outside the oven and there are official limits on the amount of microwave radiation that can be safely used. The central beam from the solar satellite will produce an amount of radiation which is far beyond these limits, so it must be carefully controlled and directed. The transmitting aerial will have some feedback device that will keep it aligned on the target, adjusting the angle with tiny puffs from the rocket engines on the satellite.

A new source of energy is needed urgently. Most of the energy we use comes from what are called fossil fuels – coal, oil and shale, in particular, but these are limited. The nations that have supplies are often in a position to ask their own price. The cost of oil has rocketed in the last decade, and the cost of coal has climbed steadily. Though the rich countries of the world complain, it is the poor countries that have been most badly hit, countries where the population is growing rapidly, while food supplies and industry are nearly static. These poor, developing countries are trapped by their poverty. They cannot set up modern industries because these need energy, and they cannot afford to buy energy at its ever increasing price. Their farmlands are unproductive because they need fertilisers and irrigation, but modern fertilisers are chemical products that need energy in their manufacture, and effective irrigation is also dependent on a ready source of energy. These poor countries are points of potential instability: their reasonable demands will eventually become so vigorous that they will threaten the stability of the rich nations.

An abundant supply of cheap energy could provide a solution for all of these problems and solar energy seems to provide the most obvious solution. The snag is that while the sunshine is free, the equipment to harness the energy isn't: if the satellite and the transmitter and the stations collecting the energy on earth last twenty years, the electricity produced has cost a twentieth of their total cost each year. And as the money to build the satellite must be borrowed and interest paid on the loan, the actual cost is considerably higher.

At present, the solar cells that convert sunshine to electricity are inefficient, making the start–up cost high. An alternative solution has been developed by the Boeing Aircraft Corporation: focus the sun's rays on a boiler in space and use the steam to drive turbines and generators. The energy would then be beamed to earth as microwaves, just as in the direct solar cell method. The idea of focusing the sun's rays isn't a new one – everyone has at some time used a magnifying glass to scorch a piece of paper in this way. In space the rays would be focused by mirrors, not by lenses: one would use a large number of flat mirrors, each positioned at an angle that would reflect sunlight on to the boiler. There are already terrestrial installations working in this way in France, in Italy and the

United States, harnessing the limited solar energy they receive.

Earthly solar reflectors have a design problem. They are heavy, yet they must be absolutely flat so that the sunlight is properly focused; this means they must be rigid. The structure must be massive, yet delicately controlled: it is therefore very expensive. The weightless mirrors in space could be flimsy – the plan is to make them of plastic, covered with a reflecting layer of aluminium – so they could be large, a couple of miles across. The Boeing project intends to use boilers and turbines running on helium, not the more conventional water.

Helium-based turbines have already been tried successfully on earth, but an enormous rocket would be needed to lift a helium turbine and some two-mile-long mirrors into space: such a rocket is still well beyond our capabilities. To make the cost anywhere near reasonable the rockets would have to be re-usable; to date, our only successes with re-usable rockets have been those used for the shuttle.

It is clear that a solar satellite cannot be built on earth and then boosted into space. The whole point about space structures is that they are flimsy and delicate, designed for a weightless world. The satellite wouldn't be able to support its own weight on earth, let alone survive being launched by a rocket, even supposing we knew how to launch objects a couple of miles in width. We could conceivably launch the satellite folded, like a giant umbrella – the solar panels on Skylab were folded for the launch – but this would be both complicated and expensive. While it would be difficult to build a satellite in space, it certainly wouldn't be impossible.

Whenever we talk of settling in space, the idea is always that the settlement will be a working one, using the special conditions – weightlessness and the lack of any atmosphere – for manufacturing processes that particularly demand such an environment. Eventually, no doubt, such settlements will be self-sufficient, but the earliest ones will be justified by the value of their exports to earth. The first export will certainly be energy.

Solar satellites will be built by people living in space, using the shuttle as a working base in a low orbit. The satellites would then be driven by rocket into their geo-stationary homes. Plans are underway for settlements that can stay up much longer than the week or so that is the shuttle's limit, and with rather more people than the half dozen or so that the shuttle can carry. A development of the Russian Salyut – itself only a tent in space – is one possibility, but first we need to establish a small settlement as a base for building a larger one that would support sufficient people to manufacture solar satellites.

The Americans already have a plan for a workshop in space which will be something like a space port. Again, Boeing has worked out a design. The living quarters comprise a couple of 50-foot cylinders attached to a docking station where space shuttles will arrive with people, food and materials. A space port is not self-contained: people will travel out to work and will have to be supported while they are there. The food and other supplies will be stored, and there will be a workshop for repairing shuttles and satellites, as well as for building solar-power satellites. It will only be a temporary base because there will be virtually no gravity. Even though it is not certain that long periods of weightlessness are harmful to the human body, the space port is not designed for long periods of residence.

Even a temporary island must solve its own energy problem. The shuttle is limited to a week or so in orbit simply because it cannot carry, or collect, enough energy for a longer period. The space port will have a large area of solar panels and use energy from these to extend its life.

There will probably be a crew of four while the space port is under trial, eight while it is working. They will live in the cylinders in a more or less normal atmosphere, and they will wear ordinary clothes. Their meals will also be normal. Most of the food will be dehydrated for travel, but there will be an abundance of water available and, once reconstituted, the food will be heated in ovens and microwave grills. A space port is no place for boiling or frying food: a raw egg could easily float away from the frying pan and drift slowly around the spacecraft.

The space-workers will need special techniques for work while they are weightless: they will stand at benches that are fitted with handgrips and their shoes will have soles that grip the ground. To sleep they will climb into sleeping-bags and drift to a sleeping area where they will anchor themselves.

The space port is a base camp in space and a storehouse. The solar-power island will be built outside the port; to work there the space-workers will have to change into spacesuits and leave the space port via an airlock. They will be able to work only for a limited time because the air supply and pressure inside the suit will be less than inside the port – well below atmospheric pressure on earth.

The space port is not very much more advanced than the shuttle – it simply needs an energy supply and extra stores – and the solar satellite is the logical next step beyond that. Both will use ideas worked out for

---

The original Boeing study of an orbiting space port for NASA. The collection of modules would be launched by space shuttle for assembly in space. Two habitat modules in the middle of the picture would serve as living quarters and contain command control, food preparation, health maintenance and recreation areas for a normal crew of eight.

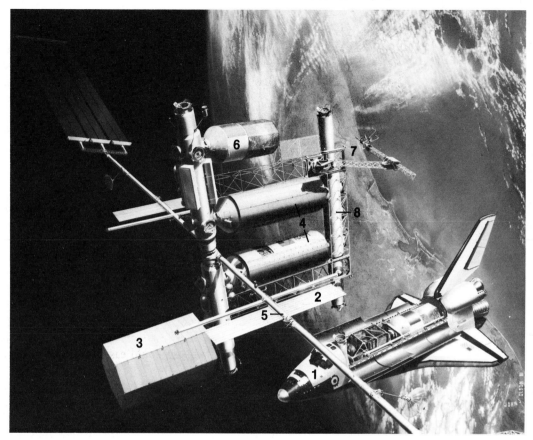

1 Space Shuttle orbiter
2 Flat panel radiator
3 Large hexagonal hangar for servicing and storing spacecraft
4 Two attached cylindrical service modules which contain propellant, batteries, power-processing units, oxygen and nitrogen
5 Long boom attached to a solar array. This array and a second at the opposite end of the boom (not shown) will provide power for the centre
6 Shorter logistics module (partially covered with thermal insulation) which serves as a storeroom for consumable items such as food, water and hydrazine
7 Track or truss structure and moveable cherry-picker crane used for handling spacecraft
8 Tubelike docking module used primarily for docking spacecraft or building other structures

the shuttle: the lightweight, working spacesuit is an obvious example. The shuttle suit is reinforced so that it can withstand the pressure difference between its inside and the vacuum of space. If the inside were kept at atmospheric pressure the suit would be far too rigid to work in, so the pressure is kept to about a third of atmospheric – 4 lb per square inch. An astronaut couldn't survive on air at this low pressure so he breathes pure oxygen which is pumped round through chemicals that remove the carbon dioxide he breathes out. The oxygen supply is sufficient for seven hours, but this includes reserves and time for coping with pressure changes.

An astronaut can't simply put on his 'low-pressure' spacesuit and leave for work. If he did, the gases dissolved in his blood at shuttle pressure would bubble off at the lower one, just as the gas bubbles off from a bottle of fizzy drink when the cap is removed. This would cause the 'bends', always a risk in space exploration: it is disabling and can kill. The astronaut must sit out a decompression period before going off to work in space.

The suit is also designed to cope with a second, slightly surprising problem. The temperature in space is very low – $-250°C$ to $-270°C$ – but that doesn't mean that an astronaut will be cold. He is in constant sunshine, wearing an impervious suit which traps his sweat. He produces a lot of heat, so the suit has a cooling system that can get rid of an average 1000 B.Th.U. per hour – twice that for a short period of very energetic activity.

The space port worker will jet to work on a remarkably versatile space manoeuvring device devised by NASA – the manned manoeuvring unit. This is rather like a hefty rucksack and contains cylinders of nitrogen attached to nozzles that act as jet engines. As the nitrogen flows through the nozzles the astronaut is driven through space. He has hand controls and can move sideways or backwards, up or down: he can even turn and spin. It sounds like an exciting toy for time off in and around the space port, but it is intended for work.

Tools used in space will have to be largely self-powered and specially designed so that using them produces no reaction. They also have to be anchored to the task or the worker so that they don't drift away.

Space construction will be made easier if the worker has somewhere to stand or sit while he works, and the Grumman Aerospace Company has already designed a 'cherry-picker' for this purpose. This is something like the cherry-pickers used for cleaning streetlamps or, come to that, for

New spacesuit and rescue system. The upper torso of the suit contains a life-support system. The rescue enclosure, a 34-inch diameter ball, contains a simplified life-support and communication system.

The Grumman Aerospace Company cherry-picker which provides the astronaut with a comfortable and safe cabin in which he can stand or sit while he works for protracted periods in space

picking cherries. It is a seat that can move out on a couple of long arms attached to the spaceport. The worker will sit in this and be carried out to the solar satellite that is being built.

The first solar satellite will no doubt be built with material brought up from earth by powerful freight rockets called space-tugs or heavy-lift launch vehicles. A space-tug can be developed from components of the shuttle. It will be automated or operated by remote control, so there is no need for a crew nor for a life-support system. The simplest plan proposed in NASA's conference on space settlement in 1975 involved replacing the aircraft-like orbiter with a simple load-carrying container. A more sophisticated design, used for carrying larger loads of up to 120 tons, puts the payload vessel on the top of the liquid-fuel container and surrounds that with four solid-fuel rockets instead of the two used on the shuttle and the simple space-tug.

Once the technique of building solar satellites has been established, space construction can become much more adventurous and much more logical. Pre-fabricating the parts of the satellite and ferrying them up by space-tug uses an enormous amount of energy, because every piece has to be lifted against the pull of the earth's gravity. This could be avoided if the parts were pre-fabricated somewhere where gravity was less powerful. An obvious candidate is the moon, which has an abundance of the minerals that are needed for construction. These could be mined and sent off to the construction site for the solar satellite – transported perhaps by a very novel piece of machinery called a mass-driver, the invention of Gerard O'Neill, one of the most stimulating planners of settlements in space. The mass-driver is a sling-shot for space, an electromagnetic catapult.

The principle is simple, though it uses some of the latest ideas in physics. The cargo – moon rock in this case – is loaded into magnetised buckets which are accelerated along an electromagnetic can or barrel by a device that switches on a series of magnets, one at a time, pulling the bucket towards the mouth of the cannon. There the bucket is stopped suddenly, but its load is carried on: the bucket returns for re-loading.

A team from Princeton University and the Massachusetts Institute of Technology has built and tested several prototypes. The bucket has to reach a very high speed so that its load can overcome the moon's gravity. It would be difficult to achieve this with ordinary magnets, so the latest design for the Princeton/MIT mass-driver uses superconducting magnets. These work on the principle that some metal alloys, when cooled to almost absolute zero ($-273°C$), suddenly lose their electrical resistance. In other words, once a current is started in a coil made of such an alloy, it goes on for ever and, because there is no electrical resistance, a very large current can be passed. The buckets can thus be turned into very powerful electromagnets. They are pulled along the barrel of the mass-driver by a series of electromagnetic coils coupled with an ingenious system of photo-electric cells which switch off the coil that the bucket has just passed and switch on the next.

Under test, the mass-driver produced an enormous acceleration of 5000 metres per second per second (i.e. 500 g), and the speed at the end of the barrel was 112 metres per second. The buckets wouldn't reach this speed if they touched the sides of the driver, so they are kept in the middle of the barrel by magnetic forces. This part of the operation is surprisingly simple: the barrel is lined with strips of copper, and the effect of the moving magnetised buckets turns them into magnets that keep the buckets in the centre of the barrel. When the mass-drivers come to be used in space they will need a more accurate guidance system. The loads will have to travel around a quarter of a million miles, carrying moon rock to a space building site where it will be collected in a massive container: a very accurate aim is needed to hit a target at that range.

The designers of the driver used an ingenious technique to halt the bucket: they wrapped the end of the barrel in a coil of wire. When the magnetised bucket reached this, it turned the wire into an electromagnet that trapped it, generating electricity at the same time. This version, known as Driver 2, has already worked well and there are plans to extend it until it eventually hurls its bucketloads out at the speed of sound on earth. It is difficult and not very useful to travel faster than on earth because of the energy needed to overcome the sound barrier – to go faster than sound. However, sound barriers exist only where there is an atmosphere: in space there is no sound barrier.

A mass-driver could also be used for 'mining' the moon. The material can at first simply be scraped from the surface – either by a bulldozer-like device or by blowing with a powerful machine that is a distant cousin of the snow-blowers used in US ski resorts – and loaded into the buckets. The line of buckets will be accelerated to reach what is called the escape velocity for the moon – the velocity at which it breaks the force of gravity – and very carefully lined up for their target, the site of the solar power station. The bucketloads will then be hurled through space and collected in the mass catcher at the building site.

The first space settlement after the spacecraft might well be on the inhospitable surface of the moon. The construction of the solar satellite will require about a million tons of moon material a year for several years. This could be mined by about a thousand people but miners won't really be space settlers – people who are going to make their lives in space. They will work in rather unpleasant conditions for spells of a year or so, living in 'huts' that are in fact cylinders much like those in the space port. The cylinders will be partly buried to insulate them from the extremes of temperature on the moon. Like the inhabitants of the space port the moon miners will live in a normal, earthly atmosphere, but work in a spacesuit supplied with pure oxygen.

The miners will need energy and solar energy may well be the answer, but it is even more difficult to harness solar energy on the moon than on earth. The moon has a day lasting fourteen of our days, followed by a night of equal length. There would have to be an efficient means of storage.

The solution could be to convert the solar energy to electricity and use that to electrolyse water into hydrogen and oxygen. These gases can be stored until needed, then they will be fed into a fuel cell, a battery that consumes the gases and converts the energy of the reaction into electricity. The hydrogen and oxygen become water in the process and will be recycled by electrolysis. All these processes have been tested and spacecraft have always had fuel cells. It is just the scale of the application that will be new. However, it is complex, and it is much more likely that in the early days a nuclear power station would be used to supply energy.

This power station will be ferried up from earth in pre-fabricated parts. We think of nuclear power stations as enormous structures, but a power station designed for use by a few people on the moon could be relatively small, though it would, of course, need better shielding than the nuclear power sources that are currently used to power orbiting satellites.

Most of the energy produced will be used to run the mass-driver – it requires a considerable amount to hurl a million tons of rock a year away from the moon, even taking its feeble gravity into consideration. But some of it will be used domestically, for cooking and air-conditioning, and some to collect the material – the ore – that is to be sent to the site for the solar satellite.

On earth these minerals would barely justify the title 'ore'. We know from the Apollo moon landings that the rocks on the surface of the moon contain aluminium, titanium and iron, all valuable metals for construction, but if they had to be smelted on earth, using energy at the price it is here, the metals would be too expensive. In space the energy for smelting is relatively freely available, and the rather increased cost can be justified by the potential value of the solar satellite.

Gradually the pattern of moon mining, the prelude to building a solar satellite, becomes clear. Though the scale is tiny in comparison with terrestrial mining operations, it is vast as a space concern. There will eventually be a thousand miners on the moon, though considerably fewer to begin with. They will have to be taken there by a shuttle taxi service and the shuttle will need more fuel for these trips than for the near-earth orbits that are its only achievement so far. So even a hundred people represents at least twenty shuttle loads. However, there would be no point in sending even a hundred people up at once, either followed or preceded by their mining equipment. There need only be a few pioneers at first, with a soil-blower and, say, one tractor. The vehicle would use fuel cells for energy.

This basecamp will be extended as more miners and equipment arrive. Over a few years the equipment to build the mass-driver and the actual mining equipment will be delivered. So will a more powerful power station and the housing for the hundred people. The probable total would be forty or fifty shuttle trips for a hundred people: the return journeys will take people back after their spell of work on the moon.

The moon matter collected by the mass-driver will be directed at the construction site for the first free-floating space community. This will slowly become a permanent self-sustaining settlement, with its own agricultural, governmental and recreational systems. There will eventually be a couple of thousand inhabitants, probably living there for a few years at a time. In a sense it will be an orbiting space factory sending energy to earth.

The space factory will be built on a bootstrap principle. A small base will be put into a near-earth orbit and peopled with, say, fifty or a hundred workers. These will gradually extend the orbiting island until it is a full-sized solar station, large enough to support its 2000 inhabitants, and available for industrial and space research.

Unlike the lunar camp or the space port, this satellite will be occupied for long periods. There must therefore be a simulated gravity in the living area. It would no doubt be possible to simulate gravity by wearing iron boots and magnetising the floor of the living area, or by combining aerodynamic footwear and a powerful wind along the floor, but these artifices and others like them would be complex solutions to the problem. The best solution is to rotate the living area. The spinning produces a centrifugal force and the inhabitants feel they are being forced outwards. This produces an artificial gravity, where 'outwards' is equivalent to 'down' on earth.

In the early days this living space would comprise about half a dozen large cylinders, about the size of the shuttle's liquid-fuel tank (154 feet long with a diameter of 27 feet). There have been plans to use surplus fuel tanks to build a space port. The cylinders would have windows, and they would be rotated in pairs at the end of an arm so that the occupants felt a pull towards the outside. They would have to be rotated rather quickly – more than once every twenty-four hours – and, although the brisk changes from light to dark would be acceptable for the early explorers, later settlers would want to find some way of achieving a more or less conventional length of day and night. The principle of this is simple: the space homes could have mirrors and shutters revolving around them to give roughly twelve hours of light, whatever the position of the rotating homes. This technique will no doubt be difficult to work out at first but it will eventually be mastered – it will have to be so that the more complex settlements can grow food.

Inside the cylinders the workers will breathe air and live on space food which is similar to the food offered in the shuttle, much like rather sophisticated camping food. Because they are living under a simulated gravity the actual living processes will be fairly normal. Work, on the other hand, will be quite unlike that on earth, if only because the island inhabitants wear spacesuits for work outside.

The first task, as ever, will be to organise an energy supply for the satellite itself. On the moon a nuclear power plant would be needed because of the length of the 'days' and 'nights'. A satellite in a near-earth orbit can have more or less continuous sunlight so a continuously running solar power station is practicable. It is certainly desirable. The first workers could build a couple of solar power stations consisting of a focusing arrangement of mirrors and a helium boiler and turbine system.

As they will eventually supply energy for a working community of 2000 people, these units will need to generate 3–4 megawatts.

The next stage will be to build the space workshops using moon rocks. It would be taking self-sufficiency too far to imagine using the first batch of lunar material to build furnaces to smelt lunar material. Most of the metal-extraction equipment and that for working the metals will have to be ferried, in pre-fabricated form, from earth, or built elsewhere in space from material ferried there.

Metallurgical experiments in space will be of particular interest. Benefiting from the ready availability of energy, the economics of production will be justified not by comparison with the cost of production on earth, but with the cost of production plus the cost of ferrying to earth. Aluminium, for example, can be extracted from anorthsite but anorthsite contains only a tiny percentage of aluminium – too little for extraction to be economically viable on earth. On the space station the process will involve smelting and electrolysis, both of which need a lot of energy. But first there should be investigation into new techniques for enrichment: it isn't really practicable, even with cheap energy, to work on very low-grade ores.

Some of the metal extraction and eventual metal working will be made easier if the plant is kept at zero gravity, but often some gravitational force will help. It would be difficult, for example, to work with molten metals in zero gravity – the liquid would disperse into droplets and drift around. If the factory workshops are situated in a rotating sphere there will be a range of simulated gravity ranging from none at all at the axis, where there is no centrifugal force, to the equivalent of gravity on earth.

Another ore that the moon will supply is ilmenite, which contains titanium. This is valuable for very strong alloys, and the liquid metal is rather easier to manipulate in space than aluminium because it is magnetic: it could be held as a molten mass by magnetic forces. This is an example of a new technique that will be possible in space, but the details are vague because we have experimented so little with manufacturing processes. When we have tested them on a small scale in, say, a space port, or even in the shuttle, we shall have a clearer idea of how to work on a large scale.

It is fairly easy to talk about building settlements and solar energy satellites in space, using raw materials from the moon; but a great deal of painstaking research and development – which is already under way – will be required to transform the theory into practice. We need to know, for example, how large a smelter can be put into orbit, and something about operating it. As working a smelter in space has its own problems – the liquid metal won't normally flow away from it – it will

189

have to be tested. The Russian Salyut stations and the US shuttles have both carried experimental smelting equipment.

The construction of solar satellites and even settlements in space is made much easier by the lack of gravity. Even very thin girders don't bend under their own weight. Thus there will be an extruder attached to the smelter to produce coils of metal ribbon that can be unreeled when they are needed and 'stitched' to form girders. The stitching could be riveting or it could be done by welding, either by an electrical process or with a weld using thermite to generate the heat that it needs. Neither process requires air. The welding will be done by computer-controlled machines, guided to the welding sites by remote control and supervised through television.

The design of the first settlement will involve tubing and spherical or pear-shaped structures of metal. These can be made in space – they can even be made on earth – by metal-spraying. The shape required is made in the form of a balloon and molten metal is sprayed on to it. In space the metal will vaporise at a low temperature and it can be squirted at a slowly rotating balloon. (We already know that we can place balloons in space – Echo 1, an early communications satellite, was a balloon made of plastic.) When the metal has hardened around it the balloon is collapsed and removed. In the very near future engineers will use the method to build the first island in space.

12

THE FIRST
SETTLEMENT

Though no one has spent much more than six months in space, and even that – a Soviet success – was a matter of camping in a satellite rather than living in it, there are plans for settlements. These will be places to stay for a long while, perhaps for life. The ideas behind these settlements do not come from vague dreamers: hard-headed scientists at NASA and in space laboratories around the world have worked out the possibilities. They see the process as a matter of 'opening up' space, as earlier explorers opened up Africa or the Western part of the United States.

The first settlers in space, like the first settlers in newly discovered parts of the earth, will be youngish men and women: children won't be taken out there. They will be adventurous and they will have to be practical people, as the early American settlers were. However, the space islands will be pleasant places, well adapted to their needs.

The island settlement itself will hover in space. It could be placed in any orbit, but a satellite in a near-earth orbit, as Skylab was, loses energy and eventually returns to earth. It would need frequent rocket-bursts to keep it up. There are locations in space, known as Lagrange points, where the gravitational effect of the moon balances that of the earth on a circulating satellite. There are three such points, known as L1, L2 and L3, on a line joining the earth and the moon, where the forces of gravitational attraction balance, but an exact balance only occurs exactly at the spot. If the satellite drifts a little towards the moon, the attraction of the moon overcomes that of the earth and the satellite continues to drift in that direction. If it drifts towards the earth, it will eventually collide with it. There are two locations off the direct joining line where the forces balance in such a way that, if the spaceship drifts slightly, the combined forces bring it back into position. These two Lagrange points are known as L4 and L5, and L5 is the one chosen for the first space settlement. There is a world-wide L5 Society of people who have dedicated themselves to being the first settlers in space.

In one of the plans formulated there would be a mass-driver on the moon. The moon's solar power satellite would be at L1 and there would be a mass-catcher at L2 for collecting material from the moon; this would rocket off to L5 whenever it had a load. Subtle calculations reveal that energy would be saved if the inhabited satellite were allowed to orbit around the L5 point rather than being controlled by rockets so that it hovered exactly there.

In the plan for moving the settlement from the near-earth orbit where it is built to L5, liquid hydrogen is first taken up to the settlement by a robot space-tug. The hydrogen would have to be refrigerated to keep it liquid, but that is not difficult in space. Then the rocket motors, which are essentially heated jets, would be taken up and attached to the settlement, rather like outboard motors. As the hydrogen is passed into the jets it expands enormously and drives the motor and the settlement forwards. The acceleration would be slow – a 2000-person settlement would reach L5 in about seven weeks, much longer than the Apollo moon trips, but quite fast enough.

The colony will need a pseudo-gravity, which could be produced by accelerating the spacecraft steadily in a straight line, but there are obvious objections to this. Every practical scheme involves rotation.

There have been various suggestions about the best shape for the rotating island. In 1929 Professor J. D. Bernal proposed a sphere, while Gerard O'Neill has suggested a rotating tubular ring, like a tyre. The rotation would generate a force throwing the inhabitants of the tyre outwards, giving the effect of gravity. It also produces the Coriolis force, which probably helps to cause space nausea. You can produce the amount of gravity required either by rotating a small space vehicle quickly or a large one more slowly, but you get much less Coriolis force, and hence less space nausea, from a slow rotation. A leisurely one revolution per minute seems to be the most that people should be asked to stand for a long time. To get the effect of normal gravity at this speed you have to be about half a mile from the axis of rotation. The best design to achieve this is a tube curved into a circle – a torus. The design worked out at a NASA conference in 1976 used a tube of 400-feet bore, bent into a circle four miles around. A torus of this size could take a population of ten thousand people. This would be a space settlement on the scale of a small town.

The structure would be wheel-shaped. The hub would have no gravity and could be used for special kinds of space manufacture and recreation. The spokes would be hollow passages from the rim to the hub. Because

---

A twenty-first-century space colony, designed to orbit between the earth and moon, as conceived by Dr Gerard K. O'Neill. This colony could accommodate a population of two hundred thousand to several million inhabitants depending on its interior plan.

193

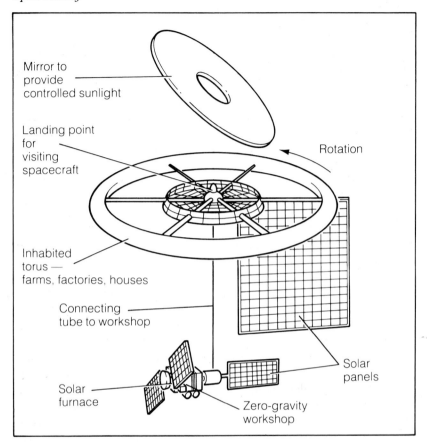

Mirror to provide controlled sunlight

Landing point for visiting spacecraft

Rotation

Inhabited torus — farms, factories, houses

Connecting tube to workshop

Solar furnace

Zero-gravity workshop

Solar panels

the fake gravity operates outwards, towards the rim of the wheel, the 'ground' on which the settlers live, build and farm would more or less correspond to the part covered by the tread on a motor tyre – around the part of the inside of the tube that is furthest from the centre. To the settlers, 'up' would be towards the centre. Travelling inwards along a spoke would feel like going upwards and need a lift; outwards along a spoke would be 'down'.

As in any satellite spinning more than once a day, the once-a-minute rotation would give some very disturbing effects of sunlight and dark, so the settlers would have an artificial twenty-four-hour cycle produced by mirrors, computer-controlled and moved by tiny thruster motors to produce the effect of a normal day-length. The sunshine is so controllable that the settlers would be able to farm very efficiently indeed.

The farming, like most of the activities of the settlement, would be self-contained. At the start everything will have to be taken to the colony

– the soil, the seeds, the atmosphere, the farm animals. These would all have to be used very efficiently as it would not be economical to ferry up such basic necessities very frequently. The farming would have to produce enough food for the settlers and any waste products would be recycled. In principle this is straightforward. The inedible parts of crops – straw, the stalks of the plants, chaff – would be used as the basis for compost, and animal manure would be used as fertiliser. It will, however, be quite difficult to make this simple idea work in practice, partly because some of the waste products will not be available at the time of the year that they are wanted, and partly because it is hard to make the amounts of any particular waste product and your need for it balance as neatly as they should in theory. This is one reason why farming on the settlement, unlike farming on earth, will have to be carefully planned. The other reason is that the demand for food can be predicted with great accuracy. It wouldn't be sensible to produce a vast crop of, say, apples when you know exactly how many potential apple-eaters there are on the space island.

On earth you cannot guarantee sunshine even in summer, but in space you can have perpetual sunshine. Equally it is possible to control the moisture and the fertiliser exactly. The only real risk of a crop failure comes from disease and to circumvent this, a reasonable excess of food will be grown. If the crop grows well, the excess will be turned into compost and recycled. While crops on earth depend on the unpredictable – weather in particular – in space we can reckon on a yield per acre which is equivalent to double the world record. The consistent sunshine will shorten the growing period for grain from 100 to 90 days. And, because there are no seasons, it will be possible to grow four crops a year. The total effect is a startling increase in production. In the American Mid-West a farmer can grow a hundred bushels of corn on an acre every year; a space farmer should be able to produce more than *four thousand*. This yield is plausible but not guaranteed – no one has yet done any space farming. One necessary project is to see just how high production can be pushed. Though no terrestrial experiments are as valuable as experiments in space they can give an idea of the possibilities, and already there are promising indications. There have been no critical experiments on growing wheat in totally controlled conditions, but there are some startling results from growing vegetables in greenhouses at Abu Dhabi. There they can produce 920 lb of tomatoes per acre *every day*, or 315 of broccoli.

Space settlers, like everyone else, may not be happy to live on grain and vegetables alone. They will probably want meat. This raises a problem in space because animals are often fed grain that could be used as food for people and, as there is only a limited amount of land in a

195

space settlement, it is wasteful to use it for growing grain to feed to animals who will convert only a small fraction of it into edible meat. However, the need for meat is more significant in space because planners must avoid anything that makes the settlement unattractive. Naturally their aim will be to use the most efficient animals. The settlers will, even so, probably have to give up some forms of meat. Cattle grow very slowly; chickens and pigs live on food that humans also eat. This may not necessarily be a problem, but it is quite likely that early space settlers will prefer to rear rabbits for meat and keep some chickens for their eggs. The rabbits could be fed on alfalfa grown especially for them and, because rabbits convert food well and breed quickly, they will provide a good return for the space that they and their food occupy. The estimate is that the settlers could produce about 145 lb of meat every day from one acre of land. They could also rear animals that live on inedible waste – outer leaves and stems. Cows and goats are the best known examples of these ultra-efficient digesters. As goats are about twice as efficient as

opposite, An artist's impression of a twenty-first century space colony and below, with landscaped interior resembling earth

196

cows in converting food into milk, it would be logical for the space
settlers to get used to drinking goats' milk.

A NASA study has found that a space settlement *could* accommodate
the whole range of meats that we are used to on earth, though it may be
impractical to produce some of them. Each settler could maintain a
'farm' of, on average, 6·2 chickens, 2·8 rabbits and a seventh of a cow.
These herds would be kept on a continuous basis: because there are no
seasons in space the animals could be replaced by breeding as they are
killed for eating. Apart from the intensiveness of the rearing this farming
would be much like that on earth. The animals would be fed intensively
and kept in the normal-gravity part of the settlement.

The study also estimates that there will be a constant supply of fish.
These could be kept in ponds, like carp on earth, but in space there
would be no need. Fish must have a wet environment and on earth they

need the buoyancy of the water, but a rotating space settlement offers a surprising alternative. At the centre – the 'hub' – of the rotating ring there is no artificial gravity and the fish could be reared there. They would float without water and their moisture need only be a cloud. There might be some problems in feeding weightless fish – you can't just sprinkle food on the surface of a cloud – but a solution could undoubtedly be found.

Space farming would make such efficient use of land that the ten thousand settlers would need only a hundred acres to provide all their food and a certain amount of personal choice would be acceptable. If they wanted cows' milk they could have it, and then they would also have veal as meat. They might have to do without beef. Their cows would be artificially inseminated with semen either stored on the settlement or sent out on one of the supply-ships. Rearing male calves to maturity, either for beef or for stud, would use rather more food than the community might wish to justify.

The space settlers would use local materials for building in designs developed from those they are familiar with, based on straight lines and right angles. Metals and glass would be easily available, and a likely design would have a framework of an aluminium alloy, clad with aluminium honeycomb for the walls. Skylab was built like this, so the technique has already been tested in space. Silicon compounds could also be used as a facing on walls – they would absorb sound and conserve energy. The ground – the curved terrain that the gravity throws the settlers towards – would be terraced, so that the buildings would be, in the settlement's sense, vertical. As there wouldn't be any rain flat roofs would be acceptable. The villages would only be subtly different from earthly ones, rather than dramatically futuristic. The elaborate system of mirrors that controls 'day' and 'night' will help to control the temperature of the houses.

The whole wheel-shaped settlement would be naturally divided by the corridor spokes and the design from the NASA conference proposed six segments, alternatively residential and agricultural. You could walk to the nearer agricultural sections from your house and there would be electrical vehicles for going further afield. If you wanted to make an extended trip you wouldn't need to travel around the circumference of the wheel: you would simply take a lift 'upwards', towards the hub, and then go down the spoke nearest your destination.

Apart from the lifts to the centre, the settlement would be a compact world with little transport – you could after all walk round it in an hour or so. The image is similar to that of a medieval town. The houses would be landscaped into hills, with the market square and the town partly concealed from view.

The design of the first settlement will offer some effects that are natural, others that certainly are not. The wheel has a four-mile circumference and the curve will be obvious to anyone looking along at ground level. The terrain doesn't have to be as flat as a Mid-Western cornfield as it is completely artificial. Someone will decide, as the planners of new towns do nowadays, that a slope or a pool or even a wood is an attractive amenity, and it will be built. Anyone who moves about the countryside around will see vistas and changes of terrain. Although the whole settlement is indoors it won't feel like it.

The views at ground level will be completely novel. The tube containing the settlement is only 400 feet thick, so the sky will be transparent and colourless. The inhabitants will be able to see across to the other side of the wheel where they will be looking at the roofs of other 'villages'. A mere two miles away there will be settlers who have an opposite sense of the directions of 'up' and 'down'.

The settlers will emigrate and live at the settlement on a permanent basis, eventually bringing up families who will become the first generation of genuine space children. And because of the permanency of the arrangement, the island has to offer the settlers a complete life. They will have a working day and they will sleep during what seems to be night in the township, although the mirrors may provide 'daylight' over the fields of growing crops. There will also be plenty of time for recreation.

There will be some unique activities. Part of the hub of the wheel could contain a low-gravity swimming-pool. There is no gravitational force at all in the exact centre, but at a distance of, say, 300 feet, the force, though small, would be enough to hold water in the pool. This would be in the form of a cylinder, with its axis the axis of the wheel. The pool water would be a coating on the inside of the cylinder. When you floated on the surface, looking upwards, you would be looking at swimmers on the other side of the cylinder, also floating and looking at you. Diving boards would be nearer the axis than the water, logically enough, but because gravity there will have only about a twentieth of the force we are used to, a diver will have time for any amount of aerobatics before a very, very gentle entry into the water. And a vigorous swimmer who pushes off from the bottom of the pool might leave the surface and float gently across the cylinder to become a diver on the other side. Space-settlement swimming will be novel and exhilarating.

Diving in the low gravity of the hub is not very different from hang-gliding on earth. Other parts of the hub's recreation area could be used for motorised gliding. The air will be too thin to give lift to a wing, but in near zero-gravity there would be no need. The motive power would come from a small rocket motor because propellers wouldn't work either. 'Space jetting' will be a safe and fascinating pursuit.

The space settlers will discover pursuits as surely as do other explorers and travellers. And in any case, some of them, at least, will want to settle in front of a screen showing TV from earth, or a 'paperback' transmitted from earth to their videoscreens.

13

WHERE DO WE GO FROM HERE?

The first settlement will be followed by others at intervals so that the later settlers will be able to benefit from the experience of the early ones. Part of the justification for building the new settlements will be space research, but the main reasons, as for the first settlement, will be intensely practical. Space settlers can build the solar energy satellites we need if our high-technology cultures are to survive the continuously growing fuel crisis. The settlements will also provide somewhere to live. We are running out of room on earth, but the growing population of the world could populate space, just as the growing population of Europe once populated America.

The material for building a series of space settlements cannot all come from the moon: it is a small satellite and, while we can easily scrape material from its surface, the idea of mining the moon, in the sense that we mine the earth, takes us into the realms of science fiction. Fortunately we now know where to find an abundance of material in space – enough, it has been calculated, to build a thousand satellite space stations the size of the earth, though no one is yet talking of space stations that large. This newly recognised source is the asteroids.

The asteroids are small rocky objects – around two thousand of them have been observed and catalogued – which travel around the sun in orbits that lie between those of Mars and Jupiter (there are a few with different orbits). They were discovered because a mathematical rule about the planets predicted that there should be a planet between Mars and Jupiter; a search for this unknown planet discovered the tiny asteroid now known as Ceres. This is the largest asteroid, with a diameter about one third of that of the moon. The asteroids are almost certainly the shattered remains of a planet that was involved in a collision in space.

We have been able to discover what the asteroids are composed of. Delicate, elegant scientific study of the light reflected from them has told us more about the minerals there than we know of some of the less accessible parts of the earth. Most of them have compositions like those

201

of the meteorites that fall to earth – they are either stony-iron or carbonaceous-chondritic. Stony-iron is self-explanatory; but the carbonaceous-chondritic meteorites, which are rare among those that fall to earth, have approximately the same composition as oil shale. Normally they burn up in the atmosphere and do not reach the earth's surface. In the asteroid belt, where there is no atmosphere, they can survive and most of the asteroids have this oil-shale composition. It is now conceivable that we will find new oil reserves in space. It will probably never be worth trying to bring the oily material back from the asteroids to earth, but it makes a lot of sense to send it to the first community at L5. In terms of the total rocket energy needed to move fuels to L5, the earth and the asteroids are about equal as sources. But we are running out of petrol-like materials on earth; we need a very powerful booster to overcome our gravity and send a spacecraft on its way to L5; and the flammable cargo must be protected as it passes through our atmosphere. There are none of these disadvantages in sending oil-shale, rich in carbon, hydrogen and nitrogen, from the asteroids to a space settlement.

First, the asteroids would be investigated by unmanned probes. These would be launched from L5 because it is so much easier to launch from a small space island with no atmosphere than from earth. Then we could start mining the asteroids.

An exhilarating possibility for an asteroid-mining craft uses what are known as solar sails. We are used to a 'brute strength' approach to space travel: we build rockets with enough thrust to get the spacecraft into orbit and then alter its orbit so that it is redirected towards our target. Once a spacecraft is actually in space, free from any strong forces of gravity, it needs only a very gentle push to move it further. Though a gentle push gives only a small acceleration the velocity will increase steadily and, if the push can be sustained, the spacecraft will in due course reach a velocity that is high enough for it to reach its target, in this case the asteroid belt, in a reasonable time. The gentle push could come from a solar sail.

The principle of the solar sail is simple. When light is reflected it imposes a pressure on the reflector which comes from the tiny 'atoms' of light, the photons. Their impact exerts a pressure on the reflector in the same way as a rain of tennis balls will exert a pressure on a wall. A photon is very tiny so the pressure from a single one is small. However, if you reflect sunlight from a gigantic sheet of foil the pressure becomes measurable. If the sheet is enormous – say a five-kilometre square – and the structure to be 'pushed' is very light and beyond the earth's atmosphere, then the pressure of the sunlight will make the reflector move away from the sun. A solar sail could in this way reach the moon in a few weeks or the belt of the asteroids in six or seven months.

And it travels free. It is consuming no fuel, which makes the economics of mining on the asteroids much more reasonable. In fact, Eric Drexler, of the Massachusetts Institute of Technology, who is the inventor and developer of the solar sail, has worked out that it might be possible even to mine common minerals and manufacture metals and alloys – iron and steel, for example – and bring them back to earth at a price that would compete with the iron and steel produced here.

The material used for the solar sail is a very thin reflecting foil, 15 to 100 millionths of a millimetre in thickness. Film as thin as this has already been made: the technique is to spread on to metal foil a substance that will evaporate when it is heated, and then deposit the reflecting film on the substance. The intermediate material is later removed by heat. On earth the whole process must take place in a vacuum which has to be elaborately created and maintained; in space there is always a vacuum.

The reflective sail is supported on a delicate framework like an umbrella, which is slowly spun to keep it from folding. The sail material is kept 'ballooned out' on its frame by the pressure of light from the sun. Drexler's plan is to launch the folded framework and the equipment for making the foil in one shuttle load – the total weight is only around 4 tons. This unit – the shuttle load – is a reasonably economic one: any project that can be contained within a single shuttle load is not unimaginably expensive.

The sails are built in panels and the equipment to do this involves a device for making sub-frames and a system for rolling out foil and coating it with its two layers – the volatile material and the thin reflector. There is nothing of great technological complexity here – the machinery that already exists for building beams in space is just as complex. There would need to be containers for the coating materials and furnaces – probably solar furnaces, with jets and pumps so that the layers could be deposited. In space it is easy to vaporise solids and deposit them accurately whereas on earth we have to create a vacuum to do it. As the panels are made, they are moved into position on the grid by crane and then fixed there. The whole process can be automatic, using the robots that industry is now developing: there is no need for human sail-builders.

Though the principle is simple, no one has yet built a solar sail in space: the equipment has still to be produced and developed. There is no way of saying precisely what solar sails will cost but they will be relatively cheap. One estimate is that the development of this sort of space technology costs around $20,000 a kilogram, or $100 million for a five-ton load.

The completed solar sails would be vast – between 0·5 and 5 kilometres in diameter, which would give a thrust of from one to a hundred newtons. The newton (N) is the metric unit of force. This force is tiny by space standards – even 100 N is only 722 poundals, whereas the shuttle launcher developed 5·3 *million* poundals from one solid-fuel engine at take-off. But solar sails do not have to overcome the gravity of earth. They will be built in orbit, about 900 kilometres up, clear of the earth's atmosphere.

The navigation of solar sails is a complicated affair, because setting sail involves tilting the spinning framework and this produces odd gyroscopic effects; however, a computer could handle it very simply. And, using only a ten-newton sail, this gently accelerating drift would take a three-ton load from the near-earth orbit, 900 kilometres up, to a geo-synchronous one, 36,000 kilometres up, in 50 days: the trip to a settlement would take longer. It isn't exactly racing into space, but it is getting there very cheaply.

Because a solar sail doesn't have to carry its fuel, long space journeys become possible. A solar sail could reach Halley's comet at its closest in 1985 in much less than a year. Solar observatories could be sent up and established in space. They could travel so far out that we would have a baseline for measuring the distance away of such mysterious objects as *quasars* or black holes. Sails could be combined with rockets to send missions to even the outermost planets. Because a sail eventually reaches a very high speed, it would take only about a year and a half for a solar-sail craft to move from a near-earth orbit to Neptune or Pluto.

The most fascinating possibilities are space prospecting, space mining and space construction. A solar sail cuts the cost of these activities so much that all sorts of projects become reasonable. With conventional rocket technology, almost the only space project that makes economic sense at present is a solar energy satellite.

Our routes to 'outer' space are – somewhat ironically – cluttered with the satellites, launchers and general debris that we have put into orbit. These objects are all plotted on charts and can be followed by radar in space. We could actually send a solar sail out there to home in on the various objects it spots, or is directed to, collect them and return them to earth, or carry them off further into space.

The most exciting possibility created by solar sails is mining the asteroids. One particular group, known as the Apollo objects, are known to be largely metallic. It would be quite simple to send a solar-sail collector to one of these: it would just have to drag a container across the surface of the asteroid, trawling up dust. The container could be large, though slow-moving, and it could pick up a tenth of a kilogram of dust each second, scraping only a few millimetres from the surface. This slow but continuous scraping would produce 200 tons in a month – a reasonable load for a 100-newton sail, which would automatically re-trim its sails and drift back to earth in about a year. The total cost of this rather

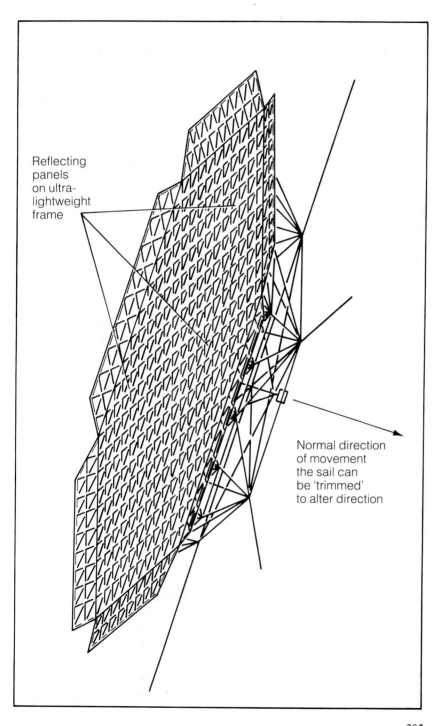

Reflecting
panels
on ultra-
lightweight
frame

Normal direction
of movement
the sail can
be 'trimmed'
to alter direction

stately operation would be no more than a couple of dollars a kilogram. And because, for a solar sail, distance costs only time, not fuel, it is sensible to mine the most valuable sources, not merely the most accessible ones. A couple of moons of Mars, for example, could turn out to be richer in potential than our own moon.

We could be using solar sails for mining by the mid-1990s. The next development which would bring the cost down further, and make asteroid mining an even better bargain, would be to build the sails from material mined in space.

The Apollo objects are rich in iron and nickel. We could use a solar-sail miner to scrape these minerals off their surfaces and bring them back to build further sails: the framework could be made of nickel steel, the reflector of nickel. The total cost of these materials, mined and refined with sails built in outer space, need be only $10 a kilogram, and the transport costs 50 cents a kilogram.

At this price we could, of course, use the materials for building space structures of all kinds. And, surprisingly, the asteroid steel would actually be cheaper than steels made from ores mined on earth. We should truly be using the resources of space to replenish the limited resources of the earth.

Because solar sails, although elegant and economical, can shift only a few tons of material each, asteroid mining will also have to use conventional methods. People would leave the L5 island in a spacecraft – a space-tug – towing space-barges to carry back the material that is mined. Once the pattern is established the space-tugs would be stationed at each end of the journey, and used to give the barges a push towards their destination. Only occasionally would the space-tug make the full journey, taking people from L5 to the asteroids or back again.

The research that has shown that it is possible to build a mass-driver that would hurl material from the moon to a mass-catcher has also shown that a mass-driver could form the basis of a very convenient form of space engine. At present we use rockets which hurl matter – gas – backwards in order to drive the spacecraft forwards. A mass-driver could be used to hurl space rubble backwards instead. The advantages would be that it is electrically powered. The space-tug could use a solar-energy system to generate electricity and harness this to operate a mass-driver. The system could generate thrust continuously as long as the supply of rubble lasted.

Mass-drivers could be used for an even more striking method of exploiting asteroid material. In a NASA study completed in 1977, Dr. Brian O'Leary worked out that it would be possible actually to capture some unusual asteroids that travel in orbits nearer the earth than the asteroid belt. The mass-drivers would simply be hitched to the asteroid and, by giving bursts of thrust at carefully calculated moments, deflect the asteroid from its orbit so that it travelled to the L5 settlement. On earth we cannot transport the mine to the miners; with this technique, it would be possible in space.

Because there is so much material available from the asteroids, and virtually unlimited energy from sunshine, there is really no limit to the number of space settlements that can be constructed. Some of these will simply be new places to move to and live in, while other settlements will be for explorers who will gradually move further and further into space. In this way a group of explorers could reach the remoter parts of the galaxy. The explorers' craft will be huge floating islands, moving slowly: the exploration will be a community endeavour.

We shall nonetheless need some new scientific developments before we entertain the idea of really remote exploration. The first series of settlements will take energy from the sun, redirecting it with mirrors. As they move further out among the planets they will need some other source of energy because they are leaving the sun behind. An approximate calculation suggests that these travellers will be able to go about ten times as far as the orbit of the most remote planet, Pluto, using known technology and, eventually, very large mirrors. But this is only a tiny distance into space. To go further the travellers will need a very compact energy store – far more powerful than anything we can use now. In his book, *The High Frontier*, Gerard O'Neill suggests using stores of matter and anti-matter which will be brought together to release energy. Anti-matter is, as its name suggests, a 'negative' form of matter. Each of the fundamental atomic particles we know – electron, proton, and so on, can have an anti-matter version – an anti-electron and an anti-proton, for example. These have been created in atomic research establishments here on earth. When a particle of 'ordinary' matter meets its anti-matter equivalent, both disappear, releasing an enormous amount of energy. We know that this occurs: we know how to react matter with anti-matter, but we know nothing about generating and storing large amounts of anti-matter.

When we remember that in twenty-five years we have progressed from the first ever artificial satellite to the space shuttle and to six-month stays in orbiting space laboratories, the next developments don't seem impossible, or even very remote. People now living will travel beyond the solar system.

GLOSSARY

Ablation The erosion or burning of a body by a hot, high-speed gas stream. Occurs to a marked extent as a spacecraft re-enters the atmosphere, and can be used to dissipate the heat of re-entry.

Abort Cut short a space launch or mission.

Absolute zero The lowest temperature possible or imaginable: it is approximately $-273 \cdot 16°C$.

Anti-matter The atoms of ordinary matter are composed of such particles as protons and electrons. Anti-matter is composed of anti-protons and anti-electrons, and although anti-matter has never been observed, these anti-particles have been artificially created. When a particle meets its anti-particle, both disappear in a burst of energy. It is assumed that the same would occur if a mass of anti-matter met a mass of matter in space.

booster The first stage of a space rocket or missile, used to initiate the flight.

coolant Liquid or gas used to transfer heat from an object.

Doppler effect Change in frequency of sound, light or radio waves caused by the movement of the source or observer relative to one another.

escape velocity The velocity needed by a particle or body to overcome the force of gravity and escape into space.

fuel cell A form of electric battery in which the substances whose chemical reaction produces the electricity are supplied continuously, so that the battery never 'runs out'.

geostationary orbit An orbit in which the satellite travels around the earth over the Equator at the same speed as the earth rotates. From the ground, the satellite appears to be stationary.

ion An atom or group of atoms that is electrically charged because it has gained or lost one or more electrons.

laser The source of a powerful beam of coherent, monochromatic light. The letters in the name stand for Light Amplification by Stimulated Emission of Radiation.

light year The distance light travels in one year: approximately six million, million miles.

module One of the self-contained units of a spacecraft: examples are the command and service modules of the Apollo craft.

photon An 'atom' of light: more precisely, a quantum of electromagnetic radiation.

photovoltaic cell A form of electric battery that converts the energy of light, usually sunlight, into electricity.

propellant The substance or substances used to provide thrust for a rocket.

solar cell A device for converting the energy of sunlight into electricity.

sounding rocket A rocket used to investigate the properties of the upper

208

atmosphere. It is launched directly upwards from the earth and falls back there.

sub-orbital   Less than orbital. Used of a space flight that travels between two points on the earth without going into orbit.

Van Allen belt   A dense zone of electrically charged particles in space around the earth discovered by the first US satellite, Explorer 1.

BIBLIOGRAPHY

Baker, David, *The History of Manned Space Flight*, New Cavendish Books, London, 1981

Clarke, Arthur C. (ed), *Coming of the Space Age*, Gollancz Ltd, London, 1967

Cooper, Henry S. F., *A House in Space*, Bantam Books, New York, 1978

Fairley, Peter, *Man on the Moon*, Arthur Barker Ltd, London, 1969

Fallaci, Oriana, *If the Sun Dies*, Collins, London, 1966

Farmer, George and Hamblin, Dora, *First on the Moon*, Michael Joseph Ltd, London, 1970

Francis, Peter, *The Planets*, Penguin Books Ltd, Harmondsworth, 1981

Gatland, Kenneth, *Robot Explorers*, Blandford, London, 1972

Johnson, Richard D. and Holbrow, Charles (ed), *Space Settlements - A Design Study*, NASA, Washington, 1977

Moller, Ron and Hartmann, William K., *The Traveller's Guide to the Solar System*, Workman Publishing Co. Inc., New York, and Macmillan London Ltd, 1981

Oberg, James E., *Red Star in Orbit*, Random House, New York, 1981

Oberg, James E., *Mission to Mars*, Stackpole, 1982

O'Neill, Gerard K., *The High Frontier*, William Morrow and Co. Inc., New York, 1977

Riabchikov, Evgeny, *Russians in Space*, Weidenfeld and Nicolson Ltd, London, 1972

Smolders, Peter L., *Soviets in Space*, English edition translated by Marian Powell, Lutterworth Press, London, 1973

Von Braun, Wernher and Ordway Frederick I., *History of Rocketry and Space Travel*, Nelson, London, 1981

Young, Hugo, Silcock, Brian and Dunn, Peter, *Journey to Tranquillity*, Jonathan Cape, London, 1969

INDEX

Page numbers in *italic* refer to the illustrations

210